# PRIMATE FIELD STUDIES

**Series Editors:**

Robert W. Sussman, Washington University—St. Louis
Natalie Vasey, Portland State University

**Series Editorial Board:**

Simon Bearder, Oxford-Brookes University
Marina Cords, Columbia University
Agustin Fuentes, Notre Dame University
Paul Garber, University of Illinois
Annie Gautier-Hion, Station Biologique de Paimpont
Joanna Lambert, University of Wisconsin
Robert D. Martin, Field Museum
Deborah Overdorff, University of Texas
Jane Phillips-Conroy, Washington University
Karen Strier, University of Wisconsin

**Series Titles:**

*The Spectral Tarsier*
Sharon L. Gursky, Texas A&M University

*Strategies of Sex and Survival in Hamadryas Baboons: Through a Female Lens*
Larissa Swedell, Queens College, The City University of New York

*The Behavioral Ecology of Callimicos and Tamarins in Northwestern Bolivia*
Leila M. Porter, The University of Washington

*The Socioecology of Adult Female Patas Monkeys and Vervets in Kenya*
Jill D. E. Pruetz, Iowa State University

*Apes of the Impenetrable Forest: The Behavioral Ecology of Sympatric Chimpanzees and Gorillas*
Craig B. Stanford, University of Southern California

*A Natural History of the Brown Mouse Lemur*
Sylvia Atsalis, Brookfield Zoo

**Forthcoming:**

*The Gibbons of Khao Yai*
Thad Q. Bartlett, The University of Texas at San Antonio

## PRIMATE FIELD STUDIES

Many of us who conduct field studies on wild primates have witnessed a decline in the venues available to publish monographic treatments of our work. As researchers we have few choices other than to publish short technical articles on discrete aspects of our work in professional journals. Also in vogue are popular expositions, often written by nonscientists. To counter this trend, we have begun this series. **Primate Field Studies** is a venue both for publishing the full complement of findings of long-term studies and for making our work accessible to a wider readership. Interested readers need not wait for atomized parts of long-term studies to be published in widely scattered journals; students need not navigate the technical literature to bring together a body of scholarship better served by being offered as a cohesive whole. We are interested in developing monographs based on single- or multi-species studies. If you wish to develop a monograph, we encourage you to contact one of the series editors.

## About the Editors:

**Robert W. Sussman** (Ph.D. Duke University) is currently Professor of Anthropology and Environmental Science at Washington University, St. Louis, Missouri, and past Editor-in-Chief of *American Anthropologist,* the flagship journal of the American Anthropological Association. His research focuses on the ecology, behavior, evolution, and conservation of nonhuman and human primates, and he has worked in Costa Rica, Guyana, Panama, Madagascar, and Mauritius. He is the author of numerous scientific publications, including *Biological Basis of Human Behavior*, Prentice Hall (1999), *Primate Ecology and Social Structure* (two volumes), Pearson Custom Publishing (2003), and *The Origin and Nature of Sociality*, Aldine de Gruyter (2004).

**Natalie Vasey** (Ph.D. Washington University) is currently Assistant Professor of Anthropology at Portland State University in Portland, Oregon. Her work explores the behavioral ecology, life history adaptations, and evolution of primates, with a focus on the endangered and recently extinct primates of Madagascar. She has presented her research at international venues and published in leading scientific journals. She is dedicated to educating students and the public-at-large about the lifestyles and conservation status of our closest relatives in the Animal Kingdom.

# The Socioecology of Adult Female Patas Monkeys and Vervets in Kenya

# The Socioecology of Adult Female Patas Monkeys and Vervets in Kenya

Jill D. E. Pruetz
*Iowa State University*

PEARSON
Prentice
Hall
Upper Saddle River, New Jersey 07458

Library of Congress Cataloging-in-Publication Data
Pruetz, Jill D. E.
  The socioecology of adult female patas monkeys and vervets in Kenya,
East Africa / J. D. Pruetz.
     p. cm. — (Primate field studies)
  Includes bibliographical references and index.
  ISBN 0-13-192787-6
  1. Patas monkey—Behavior—Kenya.  2. Cercopithecus aethiops—Behavior—Kenya.
  I. Title.
QL737.P93P78   2009
599.8'6—dc22                                                              2007051555

**Publisher:** Nancy Roberts
**Full Service Production Liaison:** Joanne Hakim
**Marketing Manager:** Lindsey Prudhomme
**Marketing Assistant:** Jessica Muraviov
**Operations Specialist:** Benjamin Smith
**Cover Art Director:** Jayne Conte
**Cover Design:** Kiwi Design
**Cover Photos:** *Main Image front cover and image on back cover*: courtesy of Jill D.E. Pruetz;
*top band images 1–4* courtesy of Robert W. Sussman/Washington University of St. Louis
**Manager, Cover Visual Research & Permissions:** Karen Sanatar
**Director, Image Resource Center:** Melinda Patelli
**Manager, Rights and Permissions:** Zina Arabia
**Manager, Visual Research:** Beth Brenzel
**Photo Coordinator:** Ang'john Ferreri
**Full-Service Project Management:** Bharath Parthasarathy
**Composition:** TexTech International
**Printer/Binder:** RR Donnelley & Sons Company

Credits and acknowledgments borrowed from other sources and reproduced, with
permission, in this textbook appear on appropriate page within text.

Pearson Education LTD., London
Pearson Education Singapore, Pte. Ltd
Pearson Education, Canada, Ltd
Pearson Education-Japan
Pearson Education Australia PTY, Limited

Pearson Education North Asia Ltd
Pearson Educación de Mexico, S.A. de C.V.
Pearson Education Malaysia, Pte. Ltd
Pearson Education, Upper Saddle River,
  New Jersey

10 9 8 7 6 5 4 3 2 1
ISBN 13: 978-0-13-192787-2
ISBN 10:    0-13-192787-6

*To Tom and my family*

# Contents

# List of Figures

# List of Tables

# Preface

The goal of this research was to examine the social behavior of female vervet and patas monkeys in relation to their ecology. The hypotheses that were tested were based on models of female primate social behavior that use patterns of food availability to predict the nature of female relationships (Isbell 1991; van Schaik 1989; Sterck et al. 1997; Wrangham 1980). Specifically, I investigated the effects of food availability on female contest competition and explored how feeding competition affected female dominance relationships in vervet and patas monkeys at the Segera Ranch study site in Laikipia, Kenya. Both patas monkeys and vervets use a whistling-thorn *Acacia* (*A. drepanolobium*) woodland habitat and vervets also use a second habitat, fever-tree *Acacia* (*A. xanthophloea*) riverine habitat. I took into account such variables as the feeding behavior of the primates, and the abundance, distribution, nutritional content, and processing costs of different food types eaten by the monkeys. Food availability was measured throughout the study, specifically concentrating on the monkeys' main food source, whistling-thorn *Acacia*. A number of measures were used to assess food availability to compare different methods as well as to assure precise measurement of primates' food sources.

Several characteristics of foods influenced patas and vervet monkey feeding competition and the dominance relationships between females. However, the frequency of contests was extremely low in both primate species. Both vervets and patas monkey females exhibited little contest competition over food in general (vervets: $N = 79$, 0.10 contests per hour; patas monkeys: $N = 101$, 0.13 contests per hour). Foods that were contested were those that were highest in nutritional quality, although abundance and distribution also influenced whether or not females would contest foods. As the models predicted, the rate of contest competition and the resultant female dominance hierarchy of female vervets was similar to patas monkeys when vervets used a typical patas monkey habitat (whistling-thorn woodland), where foods were less clumped or patchy compared to foods in the riverine habitat. The dominance hierarchy exhibited by adult female vervets when they used whistling-thorn habitat was not statistically linear, although the hierarchy order remained stable.

When vervets used the riverine habitat, where foods were clumped, the dominance hierarchy of adult females was statistically linear, as the models predicted, and females could be ranked in a sequence above or below every other female in their social group. However, food-related contest competition in this habitat alone could not explain the linearity of the hierarchy. Food-related competition resulted in a statistically linear hierarchy among female vervets only when all contests over food were considered, including those interactions that occurred in the whistling-thorn habitat, where foods were scattered. Additionally, in the riverine habitat, almost half of all agonistic interactions among female vervets ($N = 13$, or 42%) were not over food. Adult female vervets also competed over access to grooming partners, infants, and space. Although the nature of dominance relationships (i.e., linear and stable rank order) among female vervets affirmed predictions of models of female primate social behavior based on general patterns of food available to vervets at this site, food availability was not the primary factor that influenced social relationships among females. In other words, the link between ecology and behavior at the level proposed by the models of female primate social behavior (i.e., relationships) is more complex than these models maintain.

The models of female social relationships proposed by Isbell (1991), van Schaik (1989), Sterck et al. (1997), and Wrangham (1980) adequately predicted when adult female dominance relationships in patas and vervet monkeys would be nepotistic or egalitarian in a *superficial* sense. Rates of contest competition among female vervets were significantly higher in the riverine habitat (0.40 versus 0.16 contests per hour, respectively), and, a dominance hierarchy constructed using those contests was linear. However, resources other than food were contested; food-related contest competition in habitats characterized by scattered foods influenced the nature of vervet and patas monkeys' social relationships at the Segera Ranch study site; and demographic factors, in the case of patas monkeys, affected the nature of female dominance hierarchies (i.e., females entered the hierarchy yearly, and female mortality was high, resulting in high female turnover in a group). The findings of this study reveal that gross categorization of patterns of food availability are not adequate to explain female primate social behavior, and that current models primatologists use to organize behavior and ecology are too simplistic to adequately predict patterns of behavior at a number of different levels. Here, the links between food availability and female competition as well as between competition and dominance were weak. If the very basic levels of ecology and behavior are not adequately understood, the explanatory power of the current ecological and socioecological models is further reduced when attempting to correlate lifetime reproductive success with individual variation in the behaviors of interest.

**Jill D. E. Pruetz**

# Acknowledgments

This work would have not been possible without the generous permission of the Office of the President of Kenya, the direction of the National Council for Science and Technology in Kenya, and the assistance from the Institute for Primate Research (IPR) in Nairobi, especially Drs. Suleman and Bagine. My work was supported by a National Science Foundation grant (SBR 93-07477) awarded to Dr. Lynne Isbell. Roberta Fonville and Jason of Segera Ranch made my work possible via Lynne's project, and Niles Prettijohn of Segera was also of great assistance. Additionally, I owe many thanks to the Mpala Research Centre (MRC). Dr. Dan Rubenstein, of the MRC's Scientific Advisory Board, was especially helpful. John Wreford-Smith, manager of Mpala Farm, and his family were helpful to the utmost.

Many people in Kenya aided me in my effort. My work and my time in Kenya were made that much more special by the Williams family of Nairobi, including Pookie, Karl and Kathy Ammaan, Bernard Musyoka, and Julius and Sumat. The person to whom I am especially indebted is Nancy Moinde. I will forever be grateful to her and will always consider her a good friend. We had many enjoyable times together in Laikipia. Her family opened their home to me in Nairobi as well. Of course, without the guidance of and the opportunity given to me by Lynne Isbell, none of this work would have been possible. Lynne and Truman Young both helped me tremendously in my research. I have learned much from them. Researchers affiliated with the baboon project—Phillip Muruthi and colleagues, especially Valeria—provided essential companionship.

Many thanks go to my dissertation committee and to other friends and colleagues at the University of Illinois. Paul Garber provided assistance (both moral and technical!) during my research, and he, along with Steve Leigh and Stan Ambrose, contributed significantly to this work. Dr. Gene Giles helped me throughout my graduate career at the University of Illinois. I am thankful to have gained from such a knowledgeable mentor. Paul

Garber gave me my first chance to do primatological field work. Words cannot express all that I owe to Ronda Rigdon. She helped me tremendously in terms of department matters and has been a good friend as well.

A number of friends and colleagues have contributed to my achievements throughout the years. Joanna Lambert and Stephen Wooten deserve huge thanks and were wonderful friends throughout our graduate careers at the University of Illinois. David Glassman, formerly of Southwest Texas State University (and now Texas State), was my first mentor in the field of anthropology. He instilled in me an interest in primatology, whereas Dr. Jim Garber, also at Texas State, was responsible for giving me a first taste for fieldwork, in Belize. Dr. Norm Whalen of Texas State was the first person to ever suggest that I go to graduate school, and I am grateful to him for such support. Dr. Mollie Bloomsmith gave me the opportunity to work at the Michael Keeling facility in Bastrop, Texas, and I will forever be indebted to her for the excellent training that I received there.

At the center of my work is my family. Without the unfaltering support of my mother, Dorothy, my father, Orville, and my brother, Todd, I would never have gotten where I am today. The devoted support of my mother is incomparable. My family has always supported me—even when they themselves were probably wondering why I was doing what I was doing! I include Susan Lambeth in my family circle. She came to visit me while I was in Kenya, and I cannot express how special it was to share my experiences of Kenya with someone close to me. Susan has been a best friend since I met her at Bastrop, and she introduced me to the chimpanzees there. Tom LaDuke was instrumental in providing technical, academic, and moral support while I prepared my dissertation for publication as this monograph. He is truly a natural historian in the original sense and has taught me much during the years that I have known him, and I dedicate this work to him as well as to the rest of my family.

The guidance and kind invitation of Dr. R.W. Sussman, especially, as well as other editors of this special book series is gratefully acknowledged. The Prentice Hall staff, especially Publisher Nancy Roberts, have also been instrumental in aiding me to complete this work. Any and all errors found in these pages are my own.

I thank my nonhuman friends and acquaintances—Kirsten, Bandit, Pepper, C.J., Grinch, Tracey, Cochise, and Lucy—as well. Finally, in Laikipia: Taz, Burrito, Chile, Cerveza, Dahli, Georgia, Thwake, INXS, Monet, Warhol, Scooby, Elbereth, Pena, Picasso, Nietsche, Cezanne, Mooshoo, Salsa and all the others—thanks for allowing me to tag along.

# 1

# The Research Question

General ecological theory maintains that whereas male mammals are reproductively limited mainly by the number of mates available to them, a female's reproductive success is limited by the availability of foods to her (Trivers 1972; Wrangham 1980). Therefore, patterns of food distribution and abundance are likely to exert a strong influence on a female mammal's behavior. I undertook this study of vervet and patas monkeys in Kenya to test hypotheses derived from models based on these assumptions, which attempt to explain social behavior among adult female primates (Isbell 1991; van Schaik 1989; Sterck, Watts, & van Schaik 1997; Wrangham 1980). These models make predictions about the types of dominance hierarchies adult female primates will exhibit based on the distribution and abundance of foods available to them. The models assume that food availability influences competition among females and that such competition results in particular dominance styles. Feeding competition includes supplants at feeding sites or aggression during feeding (Isbell & Young 2002). Female primates are expected to compete aggressively when foods are both limited and usurpable (e.g., occurring in small patches) and despotic dominance hierarchies are predicted to emerge (Isbell 1991; van Schaik 1989; Wrangham 1980). To test the hypotheses derived from these models, I investigated the feeding behavior, social behavior, and food resources available to free-ranging vervet and patas monkeys from June 1993 to August 1995 at Segera Ranch, Laikipia, Kenya. The nature of the study site provided an excellent opportunity to examine contest competition and dominance relationships among adult females in two closely related species that use the same food resources.

## STUDY SPECIES—VERVETS AND PATAS MONKEYS

Vervets and patas monkeys are members of the tribe Cercopithecini (subfamily Cercopithecinae, family Cercopithecidae), which consists of the African guenons. Guenons are a fairly recent radiation of Old World monkeys, probably evolving within the last one million years (Leakey 1988). Mating systems of forest guenons of the genus *Cercopithecus* [this excludes vervet and patas monkeys, talapoins (*Miopithecus*), and Allen's swamp monkey (*Allenopithecus*)] range from female defense polygyny (i.e., one male monopolizes reproductive access to females) to promiscuity with multimale influxes (Cords 1988). Most guenons exhibit male intolerance of other adult males (with vervets being an exception), female defense of

Adult male vervet. *(Photo by the author)*

home ranges, and a lack of male interaction with nonreceptive females (Rowell 1988). Patas monkeys exhibit most of the aforementioned traits, whereas vervets are characterized only by female defense of home ranges.

Vervet and patas monkeys are closely related primate species (Disotell 2000; Gautier 1988; van der Kuyl et al. 1995; Martin & MacLarnon 1988; Ruvolo 1988) that have been reported to hybridize in captivity (see Lernould 1988; Matsubayashi et al. 1978). They are considered more closely related to one other than to other guenons based on chromosomal banding pattern evidence (Disotell 1996), morphological traits (Strasser & Delson 1987), and protein electrophoretic data (Ruvolo 1988). Anatomically and morphologically, vervet and patas monkeys exhibit many similarities. Each has been classified as semiterrestrial (Kingdon 1988) and as showing hand and foot adaptations geared toward terrestrial locomotion (Kingdon 1988; Strasser 1992). Patas monkeys have traditionally been classified separately from other guenon species based on such morphological characters as their long limb length.

Both patas and vervets are characterized as having overlapping home ranges with conspecific groups, aggressive intergroup interactions involving adult females, distinct mating and birthing seasons, female-resident social groups, and seasonally restricted food sources (Table 1–1). Differences between the two species include size of day and home ranges, size of within-group spread, body size, and number of adult males resident in social groups. Notable to this study, vervet and patas monkeys are characterized by

Adult male patas monkey (foreground). *(Photo by the author)*

marked differences in female-dominance relationships. Vervets exhibit a linear and stable dominance hierarchy (Cheney & Seyfarth 1990; Isbell & Pruetz 1998; Whitten 1983; Wrangham 1981), whereas dominance among female patas monkeys is more egalitarian and individualistic (Isbell &

**Table 1–1** Vervet and Patas Monkey Social Structure, Behavior, and Morphology

|  | Patas Monkeys | Vervets |
|---|---|---|
| **Social structure** | | |
| Similarities | Male dispersal | Male dispersal |
|  | Overlapping home ranges | Overlapping home ranges |
|  | Group size means: 19 ($N$ = 9), 26 ($N$ = 3), 17.8 ($N$ = 5), 36 ($N$ = 10) | Group size mean: 25 |
| Differences | Unstable female dominance hierarchy | Stable female dominance hierarchy |
|  | Non-linear female dominance hierarchy | Linear female dominance hierarchy |
|  | One-male social group | Multimale group |
|  | Home ranges not defended | Defended home ranges |
| **Behavior** | | |
| Similarities | Distinct mating season | Distinct mating season |
|  | Distinct birth season | Distinct birth season |
|  | Omnivorous diet | Omnivorous diet |
|  | *Acacia* important in diet | *Acacia* important in diet |
|  | Seasonal food resources | Seasonal food resources |
|  | Semi-terrestrial | Semi-terrestrial |
| Differences | Large day range, 4,220 m, 3,830 m | Small day range, mean 1,555 m |
|  | Large home range, mean 2,770 ha, $N$ = 2 | Small home range, 178 ha |
|  | Various sleeping sites | Regularly used sleeping sites |
|  | Patchy woodland | Continuous woodland |
| **Morphology** | | |
| Differences | Larger body size (12 kg adult male, 7 kg adult female) | Smaller body size (5.6 kg adult female, 7 kg adult male) |
|  | Longer limb length | Shorter limb length (relative to body size) |
|  | Females mature earlier (2.5 years) | Females mature later (3.5 years) |
|  | Males ~45% larger | Males ~20% larger |

Sources: *Afr J Ecol* (1994); Chism and Rowell (1986, 1988); Chism, Rowell, and Olson (1984); Fedigan and Fedigan (1988) review; Gartlan and Brain (1968); Hall (1965); Hall and Gartlan (1965); Harrison (1983, 1984, 1985); Henzi and Lucas (1980); Kaplan and Zucker (1980); Kavanagh (1978); Lee (1984); Loy and Harnois (1988); Loy et al. (1993); Nakagawa (1992, 2000); Struhsaker (1967a,b, 1969); Struhsaker and Gartlan (1970); Turner, Anapol, and Jolly (1994); Whitten (1983); Wrangham and Waterman (1981).

Adult female patas monkey. *(Photo by the author)*

Pruetz 1998; but see Nakagawa 1992). While patas monkeys are character-istically described as savanna-dwelling or savanna-woodland–dwelling primates (Chism & Rowell 1988; Hall 1965; Isbell et al. 1998; Struhsaker & Gartlan 1970), vervets are described as inhabiting riverine or gallery forest woodlands (Chism & Rowell 1988; Gartlan & Brain 1968; Pickford & Senut 1988; Struhsaker 1967b).

Vervets have been studied more extensively than patas monkeys in the wild. Long-term studies of vervets lasting at least one year in their natural habitats have been conducted in Kenya (Amboseli—Cheney & Seyfarth 1983; Struhsaker 1967a; Laikipia—Enstam & Isbell 2002; Isbell & Enstam 2002; Isbell & Pruetz 1998; Isbell, Pruetz, & Young 1998; Isbell et al. 1998; Pruetz & Isbell 2000; Samburu—Whitten 1983), Cameroon (Kavanagh 1978; Nakagawa 1991b), Senegal (Harrison 1983), and South Africa (Henzi & Lucas 1980). Patas monkeys have been studied over the long term at only a few sites. In Uganda, Hall (1965) and Struhsaker and Gartlan (1970) con-ducted the first studies of patas monkeys, but these were brief. A pro-ject undertaken by a Japanese team in Cameroon is the longest study of patas monkeys to date, but monkeys at this site are provisioned occasion-ally (Nakagawa 1989b, 1991b, 1992, 2000). Two studies of unprovisioned patas monkeys in Laikipia have been conducted, with one lasting 2 years (Mutara—Chism & Rowell 1986, 1988; Chism & Wood 1994; Chism, Rowell, & Olson 1984) and one continuing for over 10 years (Segera—Carlson & Isbell 2001; Enstam & Isbell 2002; Isbell & Enstam 2002;

Adult female vervet. *(Photo by the author)*

Isbell & Pruetz 1998; Enstam, Isbell, & De Maar 2002; Isbell, Pruetz, & Young 1998; Isbell et al. 1998; Pruetz & Isbell 2000).

## QUESTIONS AND HYPOTHESES

According to ecological and socioecological models, researchers predict that differences in female-dominance relationships will correspond with differences in patterns of food availability. In Table 1–2, I summarize what is known about vervet and patas monkey behavior relevant to the predictions of the models. I used these data to derive questions and hypotheses regarding competition and dominance in relation to patterns of food availability. Following are the questions that I addressed:

(1) How do patterns of food availability affect feeding contest competition among female vervet and patas monkeys?

(2) Are patterns of food availability a reliable indicator of the types of dominance relationships that female vervet and patas monkeys exhibit?

(3) Is feeding competition the best predictor of the style of female-dominance relationships exhibited by these primate species?

Put simply, do the differences in social behavior between female patas monkeys and vervets reflect differences in the availability of food resources

**Table 1–2** Female Social Relationships in Vervet and Patas Monkeys (Classifications Modified from Isbell 1991; van Schaik 1989; Sterck, Watts, & van Schaik 1997; Wrangham 1980)

|  | Vervets | Patas Monkeys | Sources |
|---|---|---|---|
| Social category | Resident-nepotistic (tolerant?) | Resident-egalitarian | 1, 2, 3, 5, 6 |
| Female-resident | Yes | Yes | 1, 2, 3 |
| Female relationships | Nepotistic | Individualistic? | 3, 5 |
| Within-group contests | Strong | Weak | 3 |
| Between-group contests | Yes | Yes | 1, 3, 6 |
| Dominance hierarchy | Stable, linear, despotic | Not stable or linear | 2, 3, 5 |
| Support by relatives | Common | Rare? | 3 |
| Food abundance | Limited | Limited | 6 |
| Spatial food distribution | Clumped | Dispersed | 4, 6 |
| Temporal food distribution | Long food site depletion time | Short food site depletion time | 2, 4 |

Sources: 1, Chism and Rowell (1988); 2, Whitten (1983); 3, Cheney and Seyfarth (1990); 4, Isbell and co-workers (1998b); 5, Isbell and Pruetz (1998); 6, Isbell (1991).

that each of these species exploits? The Laikipia study site provided the opportunity to directly measure the effects of food availability on feeding behavior and competition, by comparing the same study group of vervets in two different habitats and comparing vervet and patas monkeys as they use the same environment.

## SOCIOECOLOGICAL THEORY: FOOD RESOURCES AND PRIMATE BEHAVIOR

Theories of primate social behavior and ecology maintain that competition between and within social groups over access to food resources varies in response to patterns of food availability (Isbell 1991; van Schaik 1989; Sterck, Watts, & van Schaik 1997; Wrangham 1980). A major advantage of primate social grouping is that it provides certain individuals (those in social groups able to dominate other groups) with increased access to high-quality foods (Wrangham 1983). Larger groups, with more adult females, are usually dominant over smaller groups. For example, among vervets in Amboseli, Kenya, larger groups inhabited areas with a higher diversity of food plant species and with more *Acacia xanthophloea* trees, an important food at this site (Cheney & Seyfarth 1987). A potential cost of living in a group, however, is competition with group members over food resources, a critical factor to an adult female primate's reproductive success (Wrangham 1980). Females are expected to compete for foods if the benefits of obtaining them (e.g., factors contributing to greater reproductive success, such as better health) outweigh costs associated with contesting them (e.g., wounds associated with aggression). In species in which the outcome of contests between individual adult females is consistent, a straightforward dominance hierarchy emerges, and individuals can be rank ordered so that reversals in status are minimal. Where food is limited and differences in rank occur, differences in reproductive success are expected (van Schaik 1989; Wrangham 1980).

## SCRAMBLE AND CONTEST COMPETITION

Some combination of scramble and contest competition characterizes group-living primates (van Schaik 1989). Scramble competition occurs "when the net food intake of all individuals in a population is about equally affected by an increase in the population's density" (van Schaik 1989: 198) and is expressed as the adjustment of individuals' ranging behavior to group size (Isbell 1991). In other words, if animals foraging in a group have access to the same food and if all foods are considered equivalent, individuals suffer a reduction in foraging efficiency more or less equally (van Schaik 1989). These costs to foraging efficiency may be brought about, for example, by increased time and energy expended in foraging or by decreased dietary quality. Contest competition (over foods) is thought to occur whenever food distribution allows individuals to usurp high-quality food resources (van Schaik 1989). The benefits of usurping a resource, such as nutritive value gained or reduced energy costs in traveling to or searching for another resource are theoretically expected to outweigh any potential costs associated with contesting the resource (e.g., risk of wounding or reduced vigilance against predators). According to van Schaik (1989), the degree of food-related contest competition exhibited (i.e., strong or weak) is hypothesized to reflect particular patterns of food availability. Recent studies by Sussman and Garber (2004) and Sussman and co-workers (2005), however, call these assumptions into question. These authors point out the relative infrequency of social interaction between social-living primates' daily activity in general and the rarity of agonistic behavior in particular (Sussman & Garber 2004; Sussman, Garber, & Cheverud 2005).

I considered only contest competition in my study of vervets and patas monkeys, because food-related agonism is assumed to directly reflect patterns of food availability. Contest competition is a more overt measure of feeding competition than scramble competition and is less likely to be influenced by variables other than food availability (e.g., predator avoidance). Determining the monopolizability of a food has been, for the most part, a post-hoc endeavor, and authors hold differing views as to which variables best predict this quality. Theoretically, monopolizable resources include those that are distributed in space (e.g., clumped) in a manner that allows one individual to effectively keep others away. Additionally, such resources may be of high nutritive value, abundant locally but scarce generally, or reflect some combination of factors that induce one individual to usurp these foods from another. Until recently, spatial distribution was used to define the monopolizability of foods (Harcourt 1987; Whitten 1983). Clumped resources have been commonly viewed as monopolizable whereas dispersed foods are not. Isbell and co-workers (1998) suggest that primatologists have traditionally confused the importance of the spatial distribution with the temporal distribution of foods and that temporal distribution may be the most critical factor in determining resource usurpability. Clumped foods have longer food-site depletion times than dispersed foods (Isbell et al. 1998). Sterck and Steenbeek (1997) maintain that relative to social group size, medium-sized food patches are most likely to be contested because all group members cannot feed simultaneously in these patches. This prediction, however, does not take into account the ability of individuals to feed sequentially in fruit trees (Phillips 1995b) or to feed simultaneously in multiple nearby patches (Symington 1988). More precisely, contest competition should occur over medium-sized patches in relation to group size when these patches are isolated in space from other food sources. Schülke (2003) takes this into account in describing the abundance of tree gums, the main food source for fork-marked lemurs (*Phaner furcifer*) in Madagascar. The seven lemur groups' relatively small territories (1.8–6.5 ha: Schülke & Kappeler 2003) contained 14–37 of the main food trees, *Terminalia diversipilosa* or Taly tree (Schülke 2003). These trees are described as monopolizable by single individuals and relatively rare yet evenly dispersed (Schülke 2003). Such a pattern precluded group members from feeding simultaneously at the same gum site and thus presented a resource that was interpreted as invoking contest competition, as measured by physical condition of the animals (Schülke 2003).

Van Schaik (1989) predicts that any food patch that does not allow all group members to feed simultaneously promotes contest competition within groups, although both Sterck and Steenbeek (1997) and van Schaik (1989) predict that small and dispersed patches lead to reduced contest competition. Correspondingly, primarily folivorous primates are expected to exhibit lower degrees of contest competition than primarily frugivorous primates based on the general theory that leaves represent a more ubiquitous food source than fruit (but see Milton 1980). Few primates adhere to a strictly frugivorous or folivorous diet, however (see also Koenig 2002). In this study, I define monopolizable food sources as those that are no bigger than one individual's hypothesized feeding space or "personal space," which I explain later.

## SOCIAL DOMINANCE

Contest competition is hypothesized to lead to rank-related differences in feeding behavior that ultimately result in high variance in reproductive success among adult females of contrasting rank. Competition is thought to have its greatest effect on those individuals unable to secure preferred food resources (i.e., low-ranking individuals or groups). If resources are limited, low-ranking individuals are expected to experience reduced fitness through higher infant mortality, longer interbirth intervals, lower infant birth weights, and reduced health relative to higher-ranking individuals. Sade (1990) cites age at first reproduction as a critical factor contributing to differences in lifetime reproductive success among female primates; in species such as Japanese macaques (*Macaca fuscata*) and rhesus macaques (*M. mulatta*), high-ranking females matured earlier than females who are low ranking. A positive relationship between dominance rank and reproductive success is best documented for Old World monkey species. Among Hanuman langurs (*Presbytis entellus*) higher-ranking females' reproductive rate was significantly greater, and these females were more successful at preventing their infants from being kidnapped (Hrdy 1977). A 22-year study of yellow baboons (*Papio h. cynocephalus*) also showed that high-ranking females' lifetime reproductive success was significantly greater than that of low-ranking females (Wasser et al. 2004). Harcourt (1987) summarized data from twenty studies of primate species regarding the effects of dominance on adult females' reproductive success. He noted that dominant animals were likely to produce more offspring than subordinates when foods were clumped (Harcourt 1987). Whitten (1983) demonstrated such an effect for vervets in Samburu, Kenya. High-ranking females had lower rates of infant mortality than low-ranking females when their main foods (i.e., flowers of *A. tortilis*) were clumped. Whitten (1983) defined clumped foods as those that deviated significantly from random or did not correspond to a Poisson distribution, as measured by point-center-quarter (PCQ) methods. Other studies provide support for the models of female competitive behavior, but confounding factors limit their utility in this respect. For example, high-ranking, free-ranging rhesus monkey females in Puerto Rico exhibited a higher infant survival rate, and daughters of high- and middle-ranking females produced more offspring at a significantly earlier age than did lower-ranking females (Drickamer 1974). In this case, however, monkeys were provisioned, so that food was not limited compared with natural conditions, and therefore reproductive success here is correlated with dominance rank, not food availability per se. In a 13-year study of Japanese macaques, high-ranking adult females matured earlier, produced more infants, and had a shorter interbirth interval than did low-ranking females (Soumah & Yokota 1992). This population was also provisioned, blurring the association between patterns of food distribution and abundance, competition, dominance, and reproductive success.

Priority of access to resources and increased foraging efficiency are often used as indirect measures of reproductive success. Wrangham (1981) found that high-ranking female vervets successfully outcompeted lower-ranking females over access to water during times of water shortage, and low-ranking females showed significantly higher rates of mortality during this period. High rank does not always result in priority of access to resources, however. For example, among captive brown lemurs (*Eulemur fulvus*) high-ranking animals did not have better access to water during times of induced competition (Roeder & Fornasieri 1995).

## MODELS OF FEMALE SOCIAL RELATIONSHIPS

In each of the models discussed in this section, patterns of food availability are used to predict female competition to some extent, although other factors may also influence competition among primates (Figure 1–1). The terminology used to describe food availability varies. I use *availability* to refer to the distribution and abundance of foods; *distribution* is the patterning of foods in space. According to *Webster's New World Dictionary*, *distribution* is "The relative arrangement of the elements of a statistical population based on some criterion, as frequency, time, or location" (1991: 399). The terms *clumped* and *scattered* describe the spatial distribution of foods. Scattered resources may be either randomly or evenly/uniformly distributed based on statistical probability. I used the term *abundance* to

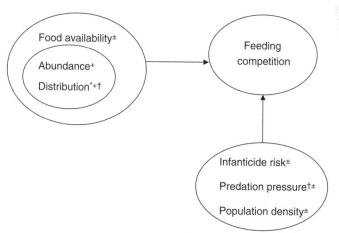

**Figure 1–1**  Factors affecting feeding competition in female primates.

*Ecological model (Wrangham 1980): food distribution = ultimate determining factor.
†Van Schaik (1989): predation = ultimate determining factor.
‡Isbell (1991) and Isbell and van Vuren (1996): food distribution and abundance = determining factor.
±Socioecological model (Sterck, Watts, & van Schaik 1997): additional factors.

refer to the quantity of food. The terms *superabundant, abundant,* and *scarce* are used to describe the relative quantity of foods.

Wrangham (1980) first described variation in female social relationships in primates according to ecological differences between species. A number of authors have since expanded his model (Isbell 1991; van Schaik 1989; Sterck, Watts, & van Schaik 1997). Wrangham (1980) used a simple dichotomy based on patterns of kinship and residence ("female-bonded" versus "non-female–bonded" species) to predict the presence or absence of dominance hierarchies. Where food is clumped, females form social groups and defend food sources from other groups. Competition within groups will occur when clumps vary in quality—for example, differing significantly in size, food type, or nutrient content. Subsequent scholars have added factors such as predation pressure, ranging behavior, infanticide, and population density to this model (Isbell 1991; van Schaik 1989; Sterck, Watts, & van Schaik 1997).

Vervets are an example of a female-bonded species, but patas monkeys are better termed a female-resident species (Isbell 1991). According to Isbell and Pruetz (1998) patas monkeys do not adhere to Wrangham's definition where "... females have highly differentiated networks of social relationships within groups, based on grooming, aggression and other interactions." (Wrangham 1980: 263). In the same article, however, Wrangham relaxed his definition to include species in which males but not females were observed to transfer between groups and those in which females were observed to breed within their natal group. According to his model, female primates live together in social groups from which they rarely or never disperse, so that they benefit from joint defense of food resources against conspecific groups. Individuals in social groups experience some level of within-group feeding competition induced by their proximity, and the intensity of such competition depends on resource availability. Competition between and within groups is thus influenced by the *distribution* of foods during differing periods of *abundance* (i.e., a "growth" compared with a "subsistence" diet). Where food is clumped, females may form social groups and defend food sources from other groups. When these resources vary in quality, within-group competition will occur, with higher-ranking females having priority access to the best food resources. During periods of food scarcity, all individuals in groups, to varying degrees, should switch to a subsistence diet that is of lower quality but is distributed uniformly (Wrangham 1980).

Van Schaik (1989) considers additional variables to explain variation in social relationships between female primates. He suggests that the type of competition females exhibit within and between social groups is influenced both by predation risk (because high predation risk forces adult females to  live in close proximity to each other) and by food distribution. He describes the patterns of competition and cooperation observed among female primates and incorporates information on dispersal habits to ultimately categorize social organization and behavior in diurnal female primates. For example, strong within-group contest competition is predicted when vulnerability to predators is high, food is clumped on a scale that does not allow all group members to feed simultaneously, and population density is less than or near that of carrying capacity (van Schaik 1989; Table 2, p. 201). Although van Schaik (1989) categorized food availability according to distribution, his definitions of clumps of food in relation to social group size actually describe food abundance because they indicate the amount of food (as well as space) available to individuals. Whereas some studies of langurs and macaques have supported van Schaik's (1989) predictions, others have not (*M. maurus*: Matsumura 1998), so further tests of the van Schaik model are necessary to evaluate its usefulness. Abundance on a small scale (i.e., per patch) is more difficult to compare across study sites than is distribution, so that its usefulness in a general model may be limited until measures of food abundance are standardized and more reliably collected.

Isbell (1991) introduced ranging behavior as a variable indicative of certain types of female social relationships. She examined the covariance of female aggression within and between groups and ranging behavior (home range size and day range length) and suggested that in species for whom food is limited, food distribution is the determining factor of within-group female social relationships. Individuals will defend (or usurp) foods when the foods are clumped in distribution (Isbell 1991). These species also tend to increase day and home range length as group size increases. Data presented in Chapman and Chapman (2000) and Sussman and Garber (2004), however, indicated no consistent relationship between group size and day range in either New World or Old World monkeys. Like Wrangham (1980), Isbell (1991) minimizes the role of predation in influencing primate sociality. She also mentions the potential effect of male reproductive strategies on females' behavior and social relationships. The model by Sterck and co-workers (1997) develops this idea.

Sterck, Watts, and van Schaik (1997) propose a socioecological model to improve upon the predictive qualities of van Schaik's (1989) ecological model. They added or emphasized factors that they suggest better explain variation in female primate behavior. They maintain that besides ecological factors, such as food distribution and abundance and predation, the risk of infanticide by males as well as pressures related to high population densities influence female social relations. Habitat saturation can exacerbate competition both within and between groups, where adult females exposed to high degree of infanticide risk could engage in cooperative defense with other females or associations with males who protect them. This model follows van Schaik (1989) and Isbell (1991) in maintaining that clumped food distribution results in strong within-group contest competition among females.

In summary, both Isbell (1991) and Sterck and co-workers (1997) maintain that clumped foods are sufficient indicators of within-group competition among adult female primates when foods are limited. Wrangham (1980) and van Schaik (1989) take into account both the abundance and distribution of foods when predicting within-group contest competition among adult females, although van Schaik labels this *distribution*. Only Wrangham's (1980) model theoretically takes into account nutritive quality or benefits gained from feeding on easy-to-process foods, referred to as *high-quality patches*'. Whereas Koenig (2002) stresses tests of the models that take an ultimate approach in providing data on correlates of fitness, such as energy gain, and on reproductive success, my study concentrates on the links central to the foundations of the models—those between resources, competition, and dominance.

## TESTING THE MODELS ON SEGERA: A NATURAL ECOLOGICAL EXPERIMENT

I tested predictions of the various models by considering food availability as an independent variable influencing feeding competition between female primates and, subsequently, dominance relationships among females within groups. I conducted my research on Segera Ranch, a commercial cattle ranch in the central plains area of the Laikipia Plateau, north-central Kenya (36°50′E, 0°15′N) (Figure 1–2). Mount Kenya (approximately 17,000

**Figure 1–2** Map showing approximate study area, Segera Ranch (filled square), Laikipia, Kenya.

feet above sea level) is 40 km to the southeast, and altitude at the study site is 1800 m (Isbell et al. 1998). Climate is seasonal, with long (April–August) and short (October–December) wet seasons and corresponding long (January–March) and short (September) dry seasons. Rainfall averages 700 mm per year but varies from year to year (Young, Stubblefield, & Isbell 1997) and even between local areas at the study site (Isbell et al. 1998). Mean monthly rainfall during my research period (July 1993–August 1995) was 48.7 mm, and rainfall totaled 565 mm for 1993 and 629 mm for 1994 (Pruetz & Isbell 2000). Months during which 60 mm or more of rain was recorded were classified as wet months, and months during which less than 60 mm of rain was recorded were classified as "dry" months (after Moore 1992).

Segera Ranch has been a privately owned conservation area since 1995 and supports over thirty species of large mammals (Isbell, Pruetz, & Young 1998; Table 1–3). Historically, black rhinoceros (*Diceros bicornis*) also inhabited this area, but they are now locally extinct. Domestic species included cattle (*Bos taurus*) and horses (*Equus caballus*). Over one hundred species of birds, including ostriches (*Struthio camelus*) and Kori bustards (*Ardeotis kori*) were also prevalent in the area.

**Table 1–3**  Some Faunal Species Resident on Segera Ranch, Laikipia, Kenya

| Order Name | Family Name | Scientific Name | Common Name |
|---|---|---|---|
| Carnivora | Canidae | *Otocyon megalotis* | Bat-eared fox |
| | Herpestidae | *Herpestes ichneumon* | Egyptian mongoose |
| Proboscidea | Elephantidae | *Loxodonta africana* | Elephant |
| Perissodactyla | Equidae | *Equus grevyi* | Grevy's zebra |
| | | *Equus burchelli* | Common zebra |
| Artiodactyla | Bovidae | *Oryx gazelle* | Oryx |
| | | *Tragelaphus oryx* | Eland |
| | | *Gazella thomsoni* | Thomson's gazelle |
| | | *Alcelaphus buselaphus* | Hartebeest |
| | | *Syncerus caffer* | Cape buffalo |
| | | *Gazella granti* | Grant's gazelle |
| | | *Aepyceros melampus* | Impala |
| | | *Sylvicapra grimmia* | Bush duiker |
| | | *Raphicerus campestris* | Steinbok |
| | | *Tragelaphus scriptus* | Bushbuck |
| | | *Redunca redunca* | Reedbuck |
| | | *Kobus ellipsiprymnus* | Waterbuck |
| | Giraffidae | *Giraffa camelopardalis* | Giraffe |
| | Suidae | *Phacochoerus aethiopicus* | Warthog |
| Tubulidentata | Orycteropidae | *Orycteropus afer* | Aardvark |

Vervets with an elephant. *(Photo by the author)*

## HOLDING HABITAT CONSTANT

A variable that was held constant in this study was habitat.[1] Both patas and vervet monkeys at the Segera study site inhabit woodland habitats. The dominant tree species in the habitat that both species share is *Acacia drepanolobium*, or whistling-thorn *Acacia*. Most of the study site consists of black cotton soils, where this tree species is found, and whistling-thorn habitat (Young, Stubblefield, & Isbell 1997). Black cotton soils are vertisols characteristic of impeded drainage (Young, Stubblefield, & Isbell 1997) that typically occur on flat land or in shallow depressions in semiarid environments (Taiti 1992). Whistling-thorn *Acacia* is the most important food species in the diet of both vervets and patas monkeys on Segera (Isbell 1998; Isbell et al. 1998; Pruetz & Isbell 2000). This *Acacia* is protected by woody thorns and by obligate ant species, which form a mutualistic relationship with the tree. The tree produces both nectaries that the ants feed on and swollen, stipular thorns that house the ants (i.e., domatia) (Hocking 1970). This food source, therefore, presents special problems for foragers,

---

[1]Corsi, de Leeuw, and Skidmore (2000) discuss what they interpret as the misuse of the word *habitat* by a number of disciplines, noting that the term has been used both as an attribute of land and as it relates to species and their biota. They recommend using a word such as *environment* to reduce confusion. Given the specific nature with which *habitat* has been used by primatologists—in reference to different vegetative types—however, I continue to use the term here.

**Figure 1–3** Author with subadult patas monkey "Taz" in whistling-thorn woodland. *(Photo by the author)*

including monkeys. Additionally, whistling-thorn *Acacia* rarely grows to heights greater than 4 or 5 m (Isbell 1998; Young, Stubblefield, & Isbell 1997), allowing the human observer to document detailed quantitative information on food availability (Figure 1–3). This presents a unique opportunity to assess, specifically, ways in which food availability affects primate behavior.

Besides the unique characteristics of whistling-thorn *Acacia*, the whistling-thorn woodland characteristic of the Segera study site provided an opportunity to test hypotheses related to the effects of ecology on primate behavior, in part, because of its lack of vegetative diversity. Young and co-workers (1997) describe the biotic community in Laikipia as one that is superficially uniform and simple. The overstory is limited to essentially one of two species of *Acacia*, of which whistling-thorn is the dominant species (Young, Stubblefield, & Isbell 1997). Using the Braun-Blanquet sociability scale of vegetation attributes, a pattern estimate places whistling-thorn at the top end of the scale: a "mostly pure population . . ." (Clarke 1986). Other woody species, as well as dominant grasses and herbs described by Young and co-workers (1997) for this site, are listed in Table 1–4. In this habitat, five grass species and two herbaceous species account for 97% of the herbaceous cover (Young, Stubblefield, & Isbell 1997).

Whistling-thorn woodland in foreground with fever-tree *Acacia* woodland behind. *(Photo by the author)*

**Table 1–4** Dominant Species in Whistling-Thorn Woodland (Young, Stubblefield, & Isbell 1997 and this study)

| Woody Species | Dominant Grasses | Dominant Herbs |
|---|---|---|
| Acacia drepanolobium | Lintonia mutans | Aerva lanata |
| A. seyal | Brachiaria lachnantha | Rhinacanthus ndorensis |
| A. girrardii | Themeda triandra | Dsychoriste radicans |
| A. mellifera | Pennisetum mezianum | Commelina sp. |
| A. brevispica | P. stramineum | |
| Balanites aegyptiaca | | |
| Cadaba farinose | | |
| Rhus natalensis | | |
| Lycium europaeum | | |

Juvenile vervet in *Acacia drepanolobium*. *(Photo by the author)*

Adult male patas monkey in *Acacia drepanolobium*. *(Photo by the author)*

Although both primate species in my study used whistling-thorn wood-land extensively, patas monkeys used this habitat exclusively whereas vervets also used a riverine environment where *Acacia xanthophloea* (fever tree) was dominant. Fever trees occur in stands along waters' edges (Isbell, Pruetz, & Young 1998; Wrangham & Waterman 1981). Riverine habitats have been characterized as having foods that are more clumped than woodland or savanna environments (Isbell, Pruetz, & Young 1998). The riverine/gallery forest at the study site borders each of two rivers that flowed through the study area: the Mutara and Suguroi rivers. Fever trees dominate this habitat and can grow as high as 25 m (Coe & Beentje 1991). Vervets' home ranges are centered along these rivers, and the study

Fever-tree *Acacia* woodland along Mutara River. *(Photo by the author)*

Juvenile patas monkeys vigilant (left) and drinking at Sugeroi River. *(Photo by the author)*

groups ranged along the Mutara River. The Sugeroi River bordered the patas monkeys' home range. A list of common woody species, herbs, and grasses characteristic of the riverine habitat are included in Table 1–5.

## OTHER FACTORS POTENTIALLY AFFECTING FEMALE SOCIAL BEHAVIOR

When examining feeding ecology, a number of variables (other than food availability and feeding behavior) that may affect competition among

**Table 1–5**  Dominant Vegetative Species in Fever Tree *Acacia* Riverine Habitat

| Woody Species | Other Species Important to the Primates Studied (i.e., Food Species) |
|---|---|
| *Acacia xanthophloea* | Grass species |
| *A. seyal* | *Cyperus* sp. |
| *A. drepanolobium* | *Brachiaria brizantha* |
| *A. girrardii* | *Pennisetum* sp. |
| *Rhus natalensis* | *Commelina* sp. |
| *Carissa edulis* | Herb species |
| *Scutia myrtina* | *Ipomoea* sp. |
| *Euclea divinorum* | *Solanum* sp. |
| | *Hibiscus* sp. |

primates should be held constant. For example, predation is thought to be a major factor influencing female primate social relationships (van Schaik 1989; Sterck, Watts, & van Schaik 1997). By living in groups, primates may improve predator detection, reduce the chances of any one individual being captured per attack, and benefit from communal defense or defense by largest group members (Cheney & Wrangham 1987). If high predation pressure leads to increased group size and group cohesion, then within-group feeding competition also may increase. On Segera Ranch, the same species of predators are common to both patas and vervet monkeys (Enstam & Isbell 2002; Isbell et al. 1998; Table 1–6). Thus, intensity of predation risk is similar for both primate species and cannot be invoked to explain differences in female social relationships. During this study, vervets on Segera were observed to give predator alarm calls in response to domestic dogs, jackals, and cheetahs, whereas patas monkeys gave alarm calls to domestic dogs, cheetahs, leopards, and lions (Isbell et al. 1998). Reptiles on Segera include puff adder (*Bitis arietans*), black mamba (*Dendroaspis polylepis*), and cobra (*Hemachatis? sp.*). Wild hunting dogs (*Lycaon pictus*) were not observed during 1992–1995, although they were considered predators of patas monkeys by Chism and Rowell (1988) at the nearby Mutara site in the late 1970s and early 1980s. Both patas monkeys and vervets avoided olive baboons (*P. h. anubis*), and olive baboons were met with alarm calls by patas monkeys (Isbell et al. 1998).

An additional factor that could theoretically affect female relationships in vervet and patas monkeys is infanticide risk. Infanticidal behavior has not been reported for vervets (Chism & Rowell 1988; Fedigan & Fedigan 1988). Enstam and co-workers (2002) reported a likely infanticidal attack by a newly resident male patas monkey in the study group several years

**Table 1–6** Potential Predators of Patas and Vervet Monkeys on Segera

| Class | Family Name | Scientific Name | Common Name |
|---|---|---|---|
| Mammalia | Felidae | *Panthera leo* | Lion |
| | | *Panthera pardus* | Leopard |
| | | *Acinonyx jubatus* | Cheetah |
| | Hyenidae | *Crocuta crocuta* | Spotted hyena |
| | | *Hyaena hyaena* | Striped hyena |
| | Canidae | *Canis mesomelas* | Black-backed jackal |
| | | *Canis adustus* | Side-striped jackal |
| | | *Canis aureus* | Golden jackal |
| | | *Canis domesticus* | Domestic dog |
| Aves | Strigidae | *Bubo lacteus* | Verraux's eagle owl |
| | Accipitridae | *Polemaetus bellicosus* | Martial eagle |
| | | *Aquila rapax* | Tawny eagle |

following my study. According to Sterck and co-workers (1997), patas monkey females might, therefore, be expected to form social bonds with one another or with males to effectively reduce the likelihood of this behavior by males. Infanticidal behavior is also thought to be characteristic of dispersal-egalitarian species (Sterck, Watts, & van Schaik 1997). My study of patas monkey social behavior, as well as other studies, suggests that this species is more egalitarian in terms of female-dominance relationships, compared with vervets, for example. The other predictions noted earlier, however, have yet to be supported by studies of patas monkey behavior. Any hypothetical threat of infanticide for patas monkeys, therefore, seems incongruent with the socioecological model of Sterck and co-workers (1997) and appears to have little influence on female patas monkey social behavior. In this study, therefore, I do not consider infanticide risk to be a variable that sufficiently explains patas monkey behavior on Segera.

Other variables that might affect female social relationships in primate species include group size and population density. Group size can affect within-group competition in primates, especially in species that exhibit stable, linear female-dominance hierarchies (Isbell 1991). For example, among brown capuchins (*C. apella*) at Manu, Peru, larger social groups exhibited significantly higher rates of within-group agonism per feeding minute than smaller groups (Janson 1988). Group size effect may be particularly relevant for savanna-dwelling species, which tend to live in larger groups and thus encounter a greater number of potential competitors (Janson & Goldsmith 1995). The groups studied here were characterized by having approximately the same number of adult females, although the patas group was larger overall than the vervet group because of the large number of immature patas monkeys. Additionally, although patas monkey females were subject to competition with one adult male during most of the year, vervet females were subject to competition with at least seven adult males during the course of this study. Both patas monkey and vervet females are subordinate to adult males in dyadic situations, so that demographic differences in group structure should be considered when assessing the competitive behavior of these females.

Population density is a variable emphasized in the socioecological model of Sterck and co-workers (1997). High population density caused by such factors as habitat fragmentation, for example, could lead to female targeting/eviction or more intense within-group competition (Sterck, Watts, & van Schaik 1997). Enstam and co-workers (2002) note that patas monkeys on Segera occur at low densities. Combined with the stable environment these primates occupy on Segera (i.e., the site is characterized as having stable populations of large mammal species and a lack of intense anthropogenic disturbance), I suggest that crowding cannot be interpreted as a factor significantly affecting female vervet or patas monkey behavior in this study.

In summary, by comparing two closely related species, I attempted to minimize phylogenetic differences that might influence patas and vervet monkeys' behavioral responses to local conditions. Differences between species could be because of ancestral relatedness rather than to the influence of food availability. For example, closely related species have generally similar diets because of anatomical traits. Thus, it is implausible to expect species that exhibit different digestive anatomy (e.g., sacculated stomachs in colobines versus simple stomachs in cercopithecines) to respond to ecological pressures similarly. Sterck and Steenbeek (1997) found that food abundance was the best indicator of langur feeding competition, but food type was the best indicator of macaque feeding competition. That langurs are less constrained by food type than macaques is not surprising, because their characteristic colobine digestive morphology enables them to exploit a lower quality diet. In attempting to determine which aspects of food availability most heavily influence primate behavior, comparing two closely related species that differ moreso in behavior than anatomy and morphology reduces confounding variables related to phylogeny. Vervet and patas monkeys are ideal subjects for comparison in that they are more closely related to each other than to other guenons (Disotell 1996), yet they differ regarding females' social relationships. In fact, the models of primate ecology and socioecology stem from data largely focusing on Old World monkey species. This taxon represents an adaptive radiation more recent than that of New World monkeys or prosimians, which may explain the support (or lack thereof) that the various taxa provide for the models of female behavior.

## PROJECT GOALS REVISITED

I examine the feeding, competitive, and dominance behavior of vervets in a typical vervet habitat type in East Africa (riverine) and compare this with their behavior in a typical patas monkey habitat type in East Africa (whistling-thorn woodland) and with patas monkey behavior. Whereas vervets are not expected to "become" patas monkeys, their behavior in the whistling-thorn woodland should be somewhat similar to patas monkeys' given that ecological variables, such as food availability, are naturally "controlled." Differences will likely exist between two species for a number of reasons (genetic differentiation, genetic drift, selection in regard to different environmental experiences: Garland & Adolph 1994). Controlling for as many variables as possible means that mistakenly interpreting difference between species is less likely. When making comparisons between species, the independent variables most often confounded are the environment and species membership (Garland & Adolph 1994). In this study, the unique whistling-thorn woodland and the vervets and patas monkeys that use it

**Table 1–7**  Subjects in Study Groups from 1993 to 1995

|  | *Patas Monkey Group* | *"Pond" Vervet Group* | *"Glade" Vervet Group* |
|---|---|---|---|
| Adult females | 14 | 9 | 3 |
| Adult males | 3 | 10 | 3 |
| Immature | >28 | 15 | 5 |
| Average group size | 40.2 | 28.6 | 9.0 |
| Range of group size | 32–45 | 26–30 | 7–9 |
| Home range size (ha) | 285 | 40 | 10 |

provided a natural experiment, and an ideal opportunity to test current models of primate behavior.

## Study Groups

Study groups included two groups of vervets and one group of patas monkeys (Table 1–7). Patas and vervet monkey study groups did not have overlapping home ranges but were located within a distance of approximately 4 km. The vervet study groups had adjacent, overlapping home ranges, whereas the patas monkey group's home range overlapped with a group of vervets not included in this study because of its periodic crop-raiding behavior. The vervet groups' ranges also overlapped with the home range of one troop of baboons. The patas monkey study group overlapped with this baboon troop and perhaps with a second. Lesser bushbabies (*Galago senegalensis*) shared home ranges with all study groups. Each vervet group's range overlapped with a conspecific group at opposite ends of their respective home ranges. The patas monkey study group's home range overlapped with at least two other patas groups.

## Study Subjects

Study subjects were adult females in one patas monkey and two vervet groups (Pond and Glade groups), although certain data were collected on the feeding and contest behavior of adult males and immature individuals in all groups (see Pruetz 1999). Data on the Glade group were collected for comparison in terms of feeding behavior and contest competition but were not included in analyses of dominance because the small number of adult females precluded statistical tests of linearity (Appleby 1983). Ecological data for vervets were collected in the overlap zone between the two focal vervet groups.

Most monkeys were individually identifiable. All vervets were recognizable based on wounds, scars, pelage, and morphological differences.

Adults and, occasionally, subadult (<2 years of age) patas monkeys were individually recognizable based on pelage and morphological differences, such as nipple coloration and length, ischial callosity coloration, body size, and hair color. Immature patas monkeys were identifiable only according to age (infant, 1-year old, 2-year old, 3-year old). Maternal kinship was known for vervet infants and for 1-year- and 2-year-old juvenile vervets. Group members were habituated to a distance of 10–20 m and often to 5 m. Habituation of study groups had begun in June 1992 when Lynne Isbell initiated her long-term research project on the study groups. Her long-term project on vervet and patas monkeys and the unique nature of the site gave me the opportunity to examine the feeding ecology of these species to such a degree that is not often possible in many field situations.

# 2

# Measuring Primate Behavior and Ecology

The goal of my research was to examine the strength of the relationship between primate social behavior and ecology as predicted by socioecological models put forth in recent years (Isbell 1991; van Schaik 1989; Sterck, Watts, & van Schaik 1997; Wrangham 1980). In order to do this, I began by quantifying data on food availability for patas monkeys and vervets. At the core of each of the socioecological models lies the assumption that patterns of food availability significantly influence a primate's behavior. Scattered, abundant, or otherwise not usurpable food resources are predicted to result in adult female-dominance hierarchies that are less stable and nonlinear, with a high percentage of rank reversals (Isbell & Pruetz 1998; Figure 2–1). If food availability has an important effect on behaviors such as intra-group dominance relations, comparing closely related species that exploit similar resources can help to establish the effects of environmental variables on social behavior and ecology. Whereas a central theory in the study of primate socioecology is that the environment and, especially, food resources within the environment influence primate social structure and behavior (Barton, Byrne, & Whiten 1996; Chapman, White, & Wrangham 1994; Crook & Gartlan 1966; Doran & McNeilage 1998; Isbell & van Vuren 1996; Nakagawa 1998; van Schaik & van Hooff 1996; Struhsaker 1969; Suzuki, Noma, & Izawa 1998; Temerin & Cant 1983; Watts 1996; White 1996; Wrangham 1986; Wrangham et al. 1996; Yeager & Kirkpatrick 1998), accurate measures of the food available to primates are often lacking. Researchers have frequently relied on qualitative estimates and gross measures of food availability. Using general measures as well as the lack of

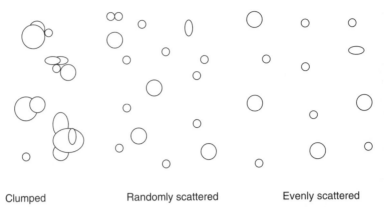

| Clumped | Randomly scattered | Evenly scattered |

**Figure 2–1** Categories used to describe food distribution in space (after Terborgh 1992: 85).

standardization between studies has limited our ability to identify precise relationships between the availability of primate foods and behavior such as feeding competition. Some researchers, however, have attempted to remedy this situation by providing guidelines for between-site comparisons (see Janson & Chapman 1999; edited volume by Kuroda & Tutin 1993). In general, ecological studies of primates have been lacking in (1) accurate quantification of food availability and (2) comparability of ecological data across sites and studies. The Segera study site in Kenya allowed environment to be held constant when assessing the effects of food resources on social behavior and dominance relationships among adult females of closely related species, while providing a situation especially conducive to ecological study.

## "MEASURING" ECOLOGY

Detailed study of an organism's environment is essential for the proper interpretation of its behavior. For example, using knowledge about ecological influences on living primates' behavior to reconstruct the lives of extinct primate species, including humans, is central to the discipline of biological anthropology. The accurate interpretation of the fossil record is challenged by a number of issues, such as accurately dating remains and reconstructing the environment in which fossil species lived. Using qualitative rather than quantitative measures of primate food availability to interpret the socioecology of extinct species merely adds to the difficulties associated with the study of fossil species. For instance, the evolution of a fission–fusion social system and the long day ranges of Hamadryas baboons (*P. hamadryas hamadryas*) have been interpreted as ancestral adaptations to low density resources, such as sleeping cliffs and food resources

(Kummer 1990). Social patterns such as male herding, male "ownership" of females, and male bonding are thought to have evolved in a habitat where resources (e.g., sleeping cliffs) were large enough to permit the aggregation of large troops, but where food resources were distributed in a way as to select for fusion into bands by one-male units (OMUs) and clans (Kummer 1990). In this case, sleeping cliffs are an example of a single, large, shareable resource that offers protection for hundreds of individuals. Extant Hamadryas baboon OMUs congregate into clans and bands to share sleeping cliffs. In contrast, resources exploited by OMUs (e.g., food) are characterized as "sparse," limiting the number of individuals that can forage together (Kummer 1990). Kummer (1990) notes, however, that quantitative data on the availability of food resources are lacking. Thus, the latter component of this hypothesis has yet to be tested. This example is not atypical of theories that propose to reconstruct the behavior of extinct primate species. We often use hypothesized rather than quantified patterns of food availability to explain extant primate behavior and subsequently apply these assumptions to extinct species.

Whereas many studies of nonhuman primates have demonstrated that ecological factors, such as food availability, affect aspects of primate behavior (Budnitz 1978; Crook & Gartlan 1966; Gartlan & Brain 1968; Isbell, Pruetz, & Young 1998; Muruthi, Altmann, & Altmann 1991; van Schaik 1989; Wrangham 1986), a better understanding of general ecological principles is hampered by a lack of standardization between field studies (see also Janson & Chapman 1999). Consider two different studies of vervets. Michael Harrison (1984) found that vervets in Senegal were more selective when foods important in the diet were common. He quantified food availability by recording the presence or absence of vervet foods in quadrats that comprised the monkeys' home range, randomly choosing a subsample in which to count all trees of important food species. The height and average food abundance of fifteen trees of each species were then measured to estimate average abundance of these species, whereas fifty trees per month were randomly selected and sampled regarding the presence or absence of different food types (i.e., phenology). Harrison (1984) then calculated overall food availability using a formula that included data on plant distribution, density, size, and phenology. In a second study of vervets, in Kenya, Patricia Whitten (1983) found that differences in diet among females of different ranks were apparent for clumped foods but not for foods that were randomly distributed. Food availability for vervets in her study was assessed using both general (e.g., scoring the mean percentage cover by food species in different vegetation communities in the vervets' home range) and more specific measures (Whitten 1982, 1983). For example, she used point-center-quarter (PCQ) measures of food species density and distribution in each study group's home range. She collected data on three different feeding tree species, and rated them on a

5-point percentage scale as to the relative availability of different vervet food types, such as flowers and fruits. Diameter at breast height (DBH) and crown diameter were recorded for these trees (Whitten 1982). Grasses were scored according to their greenness, and forbs were scored according to their fruit and/or flowering condition on a 5-point scale. Finally, Whitten (1982) counted the number of individuals of each feeding-tree species within the vervets' home range, as well as noting the reproductive state of these trees. Whereas both Whitten and Harrison employed useful measures of food availability, data on food abundance within patches (i.e., presence or absence of foods within tree crowns versus the amount of these foods scored on a percentage scale) at these sites are not comparable. If one wanted to address the question of intersite differences, data on food distribution (densities using quadrats versus PCQ measures) are comparable, however, because these data were collected using standardized scores or measures (e.g., meters). Different methods may reflect the questions addressed and hypotheses tested, but standardization of methods allows for more accurate interpretations of adaptive patterns.

An assumption made by many primatologists is that food availability varies greatly according to season. Many studies have shown seasonal changes in behavior throughout the order Primates. Seasonal changes in primate behavior, such as time spent eating, processing, and foraging, are usually assumed to reflect seasonal changes in the availability of foods. Differences in food abundance between seasons, though, have often been only qualitatively described (e.g., Japanese macaques: Agetsuma 1995a). Johnson (1990) found that adult male yellow baboons (*P. h. cynocephalus*) spent less time feeding on foods with relatively high fiber content in the dry season but exhibited no such selectivity during the wet season. She assumed that foods available during the dry season were generally higher in fiber, but no data were given to support this conclusion (Johnson 1990). Others have linked changes in primate behavior with more quantitative evidence of seasonal differences in food availability. For example, Byrne and co-workers (1990) attempted to compare food availability across groups and across sites for different study groups and species of baboons (*Papio* sp., *Hamadryas* sp.). These authors used the nutritional quality of all baboon foods to infer the quality of diet available to the different groups and used strip plots (i.e., quadrats along line transects) of differing dimensions to measure the densities of underground storage organs (USO), herbaceous vegetation, and tree resources included in the diet. However, foods were only measured when baboons were observed to feed on them and not at times when these same foods may have been available but were not included in the diet. Using data on feeding behavior, nutritive quality of foods, and density of food plants, Byrne and co-workers (1990) concluded that baboons switched from foods hard to find and process (e.g., deeply buried USO) to those easy to find and process (e.g., tree fruits) during seasons when the latter were more readily available (i.e., spring/rainy season). These authors do point out that not all baboon foods were recorded using the strip plot method and that, for highly selective feeders such as baboons, this presents a severe limitation on the accurate assessment of food availability.

An example of the utility of using quantitative measures of food availability stems from tests of the terrestrial herbaceous vegetation (THV) hypothesis, proposed to explain the differences observed between chimpanzees and bonobos. Wrangham (1986) first hypothesized that the THV used by bonobos was a consistently abundant and evenly distributed food source compared with the tree-fruits characteristically used by chimpanzees. He attributed the larger mean group-sizes of bonobos relative to those of chimpanzees to relaxed levels of feeding competition in large THV food patches. Wrangham and colleagues did not quantify the abundance and distribution of foods, however, until later studies were conducted. Using DBH, crown radius measures, and the density of food trees available to bonobos and chimpanzees along transects, Chapman and co-workers (1994) found that the size and density of tree-fruit patches were similar for bonobos at Lomako and for chimpanzees in Kibale Forest, Uganda. Chimpanzees at Kibale also fed more on THV (12% of diet) than did bonobos at Lomako (2% of diet). This latter finding contradicted Wrangham's (1986) assumption of the importance of THV to the diet of bonobos. Chapman and co-workers (1994) suggested that the spatial distribution of foods did not contribute to the observed differences in chimpanzee and bonobo party sizes, although the *distribution* of bonobo and chimpanzee foods was not quantified. The authors concluded that temporal differences in food availability may be responsible for the differences in social structure and behavior exhibited by chimpanzees and bonobos, using Malenky's (1990) data on seasonal fluctuations in rainfall and his general measures of the amount of fruit produced each month (species fruiting per month) as evidence. More recently, Wrangham and co-workers (1996) have used the quality of THV to explain the differences in grouping behavior between chimpanzees and bonobos, hypothesizing that bonobos are able to form larger groups because they have greater access to protein-rich high-quality THV, whereas chimpanzees do not. This series of studies illustrates the importance of quantifying a number of aspects of food availability to better address questions of primate socioecology.

## MEASURING FOOD AVAILABILITY

### Commonly Used Methods

That assessing primate food availability is difficult is generally acknowledged (Byrne, Whiten, & Henzi 1990; Chapman 1988; Isbell & Pruetz 1998;

Janson & Chapman 1999; Malenky et al. 1993; Oates 1987; Wrangham 1980). Occasionally, interpretations regarding the effects of food availability on primate behavior and ecology are based on qualitative data (e.g., Barton et al. 1992; Ihobe 1989; Samuels & Altmann 1991; Yamagiwa et al. 1992) or use characteristics of primates' diets as an indicator of food availability (e.g., Boesch 1996; Fossey & Harcourt 1977; Matsumoto-Oda et al. 1998; Wrangham 1977). In a study of the diet and feeding behavior of vervets in Cameroon, Kavanagh (1978) used qualitative descriptions of food availability as evidence that vervet foods were clumped to some degree, in turn using these patterns as an explanation of why the monkeys engaged in little search time at feeding sites. The absence of aggressive disputes over foods, however, was assumed to be because of the fact that even during periods of supposed food shortage (i.e., dry season), foods were not "sufficiently clumped to be defended" (Kavanagh 1978: 40). "Clumped" is used as a relative term to describe the author's objective perception of patterns of food availability at two different scales, and no quantitative data are given to support these conclusions. Most studies of primates, however, rely minimally on some quantifiable indicator or measure of food availability, such as rainfall or food plant dimensions.

Many studies of free-ranging primates have inferred food availability from the amount of rainfall, especially in areas that exhibit marked dry and wet seasons (see Popp 1983). For example, adult female yellow baboons were reported to have conceived "after months of heavy rain when food was relatively abundant" (Silk 1987: 611) and to have "spent progressively more time feeding and less time grooming than expected . . ." based upon the amount of recent rainfall (593). Similarly, Henzi and co-workers (1997) assumed that foods would be scarcer for savanna baboons during the dry season and also assumed homogeneity of food availability in the baboons' range based on the existence of a "single floristic belt." Doran (1997) and Boesch (1996) each noted the scarcity of resources available to Tai Forest chimpanzees. They based their interpretation on the qualitative assessment that the major dry season preceding the study was "extremely severe . . . and that the year as a whole had much scarcer resources than usual" (Doran 1997: 185), in part because almost no *Coula* nuts had been produced (Boesch 1996). Even though such assumptions may be correct (see Barton et al. 1992; but also see Watts 1998), more detailed quantification of foods is sorely needed to understand primate behavior and ecology more specifically. An ongoing study of savanna chimpanzees (*P. t. verus*) in Senegal reveals that late dry season months are those that are highest in fruit availability (Pruetz, 2006). Nakagawa (2000) also found this to be the case for patas monkeys in Cameroon.

Most studies of primate food availability have used indirect measures (e.g., DBH, crown volume, tree height, relative phenology scores) and focus on major food species. For example, Schülke (2003) used transects to estimate the abundance of the ten most important foods to fork-marked lemurs and found that neither DBH nor crown volume was as good an indicator of food availability as estimates using counts of food items. To determine which food characteristics were instrumental in fruit selection by woolly monkeys (*Lagothrix lagothrica*), Stevenson (2004) considered such factors as DBH, plant density, crop duration (a temporal measure), and numerous phytochemical properties as well as fruit dimensions on a small scale. Abundance at a large scale combined with fruit astringency was most important in this case (Stevenson 2004). On a smaller scale, commonly used but indirect methods of quantifying food availability include the use of fruit traps (Garber 1988; Goldizen et al. 1988; Malenky et al. 1993; Terborgh 1983; Zhang 1995), fruit trails (Guillotin, Dubost, & Sabatier 1994; Rogers et al. 1998; White 1998), or, in the case of macaques, seed traps (Nakagawa 1991a). Food items falling into such traps or along a specified trail can be quantified in several different ways—for example, counting, weighing, and measuring. Bias against foods that are consumed is inherent when using this method. For example, fruits that are preferred by frugivores, those that are slow ripening so that they are eaten in greater proportion than fast-ripening fruits, and fruits consumed in greater proportion during periods of fruit scarcity may be underrepresented in fruit traps (Malenky et al. 1993). Additionally, especially in the case of fruits, parasitized items may be overrepresented when, in fact, primates would reject such items (although they may also selectively choose them). Fruits that have been aborted by the parent tree are likely to be overrepresented in fruit traps (Malenky et al. 1993). In light of these points, considering primates' food preferences when assessing food availability is especially important (Malenky et al. 1993).

Increasingly more field researchers use a variety of methods to evaluate the availability of foods to primates and/or assess the reliability of the measures used (see Stevenson 2004 example above). For example, Koenig and co-workers (1998) used measures of basal girth, girth at breast height, two crown diameters, and plant height to assess the abundance of foods available to langurs (*P. entellus*) at Ramnagar, Nepal. In analyses of feeding party size, however, which was hypothesized as related to food availability, only tree height was used as an indicator of food availability (Koenig et al. 1998). Whether this particular variable was the best predictor of food abundance to langurs is unclear. Phillips (1995b) assessed the utility of DBH in estimating crown volume of fruiting trees fed in by white-faced capuchins (*Cebus capucinus*) by comparing this measure with crown volume for a subset of feeding trees. Even though DBH was an accurate predictor of crown volume, the accuracy with which crown volume can be used as an indicator of food abundance for different tree species is unknown. In a study comparing feeding time by golden lion tamarins (*Leontopithecus rosalia*) on fruits to various measures of availability, Miller and Dietz (2004) found that fruit yield as measured in dry grams of fruit matter per fruit-bearing region was more highly correlated with tamarin feeding time than was DBH or fruit crown volume. In a study of the abundance of foods available to Kibale Forest chimpanzees, Malenky and co-workers (1993) noted the costs and benefits of using different measures (fruit traps, phenology transects, and trails) but could not assess the accuracy of the different methods because data on actual food abundance were not attainable. These authors, however, were able to make recommendations on the feasibility of using the different measures based on available time and the potential biases of each method (Malenky et al. 1993). Using a number of measures and assessing the usefulness of each is a step toward more accurate quantification of primate food availability.

Counts of food items are a more accurate measure of food availability. In a test of methods, Budnitz (1978) found no difference in the availability of a main food species (*Tamarindus indica*) in different ring-tailed lemur (*Lemur catta*) habitats using direct measurements of fruit production (counts of fallen pods) but found significant differences according to habitat using indirect measures (crown size, height, DBH). However, direct measurements of food availability were limited because only foods that had fallen to the ground were quantified (Budnitz 1978). Some studies on cercopithecids provide data on nearly absolute (not estimated) food availability (Agetsuma 1995b; Barton et al. 1992; Hamilton, Buskirk, & Buskirk 1976; Marsh 1993; Suzuki, Noma, & Izawa 1998; Watanuki et al. 1994). Hamilton and co-workers (1976) provided counts of the number of seedpods in trees and on the ground for two *Acacia* species that were a major component of the diet of chacma baboons (*P. h. ursinus*) in the Republic of South Africa. Agetsuma (1995b) similarly quantified fallen seeds for Japanese macaques at Yakushima. Studies such as one by Watanuki and co-workers (1994) combine counts with estimates based on a subsample. These authors quantified the number of mulberry tree buds available to Japanese macaques at Shimokita Peninsula by counting the number of shoots on seventy trees, measuring the length of shoots and number of buds per shoot on thirty shoots, and using these data to estimate the number of buds per tree on trees for which shoots had been counted. In contrast, some studies have estimated counts of foods in categories of anywhere from hundreds to thousands of food items (Post 1987; Sterck & Steenbeek 1997; Utami et al. 1997). In one of the few examples where the absolute abundance of food has been quantified in primary forest habitat, Zhang (1995) used platform observation to provide counts of fruits available to brown capuchins in French Guiana. The number of fruits within a 20 m radius from platforms and walkways between platforms was counted using binoculars (Zhang 1995). The use of absolute counts is an accurate measure of primate food availability, although logistical problems, such as time and workforce constraints, may prohibit the use of such measures.

Perhaps the most precise assessment of food availability is to quantify foods according to the number of bites available to focal subjects and the energy derived from this unit of measurement. Byrne and co-workers (1990) quantified the number of potential bites available to chacma baboons in South Africa for tree foods, herb foods, and USOs. Barton and co-workers (1992) used such methods to assess the habitat use of olive baboons in Laikipia, Kenya, and concluded that the distribution of foods rather than their abundance influenced baboons' ranging patterns. It was suggested that rather than an overall lower biomass during periods of food scarcity (i.e., dry season), baboons were faced with increased patchiness of foods at such times. This could explain baboons' increased patch residency times during periods of decreased food availability, but these authors failed to quantify the size of food patches used by baboons. Nonetheless, this method provides a very accurate measure of primate food abundance on a grand scale, as well as of primates' perception of their food resources, and one that is potentially widely applicable for comparison across study sites. Nakagawa (2000) used this method to assess the daily intake of available energy and gross protein for patas and tantalus monkeys (*Cercopithecus aethiops tantalus*) in Cameroon. He was, however, limited to sampling only one subject in each age-sex class because of the intensity of this type of sampling.

Behavioral methods have also been used to infer the availability of foods to primates. Feeding and ranging behavior has been used to indirectly assess the patterns of food abundance and distribution for free-ranging vervet and patas monkeys (Isbell et al. 1998). The movement of adult females between feeding sites and the average time spent per site was used as an indicator of the distribution and abundance of foods, respectively. Similarly, Iwamoto (1992) used the ranging behavior of Japanese macaques as an indirect measure of food availability. The distribution of feeding sites and the length of feeding bouts within marked $50 \text{ m}^2$ quadrats were taken to represent the distribution and abundance of these primates' foods, respectively (Iwamoto 1992). Garber (1986), using behavioral data on moustached tamarins (*Saguinus mystax*) and saddle-backed tamarins (*S. fuscicollis*), concluded that these species' food sources produced small amounts of fruit per day. He based his interpretations on the observation that tamarins used many feeding trees per day, visited trees once a day, and visited the same trees on successive days (Garber 1986). Recently, Williams and co-workers (2004) used chimpanzee party size as a measure of overall food availability, although other factors such as number of estrus females may be equally influential to chimpanzee party size (Piel 2004). Combining behavioral measures with exact measures of food availability (e.g., counts of foods per plant species, plant species, number of bites available) may be the best technique for understanding the relationships between primates and their environment because an objective assessment of the foods available to primates can be combined with their own interpretation of these resources.

### Primates' Perceptions of the Foods Available to Them

Determining the appropriate scale at which to measure food availability will depend on the hypotheses being tested, but measures of food availability should ideally take into account how the study subject perceives the available food resources. An animal's age, sex, nutrient requirements, and body size are variables that may influence an individual's perception of its environment. Generally, defining a food "patch" according to the relevant questions one wants to address is problematic (Byrne, Whiten, & Henzi 1990; Chapman 1988; Chapman, White, & Wrangham 1994; Isabirye-Basuta 1988; Isbell et al. 1998; Nakagawa 1989a; Oates 1987; Symington 1988). Foods can be described in relation to individuals within social groups or communities or in relation to entire social groups or communities. For example, patches of foods varying in size from small to medium have been predicted to incite competition within a social group (van Schaik 1989; Sterck, Watts, & van Schaik 1997; Symington 1988), whereas large patches of food that allow all group members to feed simultaneously are expected to result in competition between but not within social groups (Symington 1988; Wrangham 1980). Traditionally, individual feeding-tree crowns have been referred to as a single patch (Garber 1997; Nakagawa 1989a; Phillips 1995b; Sterck & Steenbeek 1997). *A. xanthophloea* trees, however, were described as patchy resources for yellow baboons in Amboseli, Kenya, based on the distribution of the habitat types in which these trees were found (Post 1982). White and Wrangham (1988) and Chapman (1988) defined a *patch* as an area in which an individual can continuously move and eat. This definition takes into account individual differences, such as body size, that may influence a primate's perception of a patch and would be best suited for studies of how individuals rather than entire social groups perceive food resources. A study of olive baboons at a Laikipia, Kenya site designated single *A. tortilis* trees as the most clearly definable patches (Barton 1993). Describing patch size relative to social group size or relative to the individual and providing data on the absolute sizes of these patches would more effectively reveal the effects of food availability on primate behavior than if either of these methods were considered alone. The terms "patch" and "patchy" should be used to refer to size and distribution, respectively.

Studying primates that commonly use foods not easily recognized by human observers as defendable patches presents additional obstacles to the assessment of food availability. Most primates include some foods in their diet that do not occur in easily definable patches, e.g., non-social insects. Although savanna or woodland-savanna habitats do not present the same difficulties associated with measuring food availability as do forested habitats (e.g., access to food items such as fruits within crowns of tall trees), other difficulties emerge. Byrne and co-workers (1990) point out that because most patches in forested habitats are individual tree crowns, we can compare food availability using standardized measures. Thus, the problems associated with measuring food availability in a forest habitat are the same for most foods in such an environment (Byrne, Whiten, & Henzi 1990). Foods available to savanna and savanna-dwelling primates such as baboons (and patas and vervet monkeys, as well as chimpanzees at some sites), however, are distributed on extremely varying scales (Barton et al. 1992; Byrne, Whiten, & Henzi 1990; Shopland 1987). Temerin and Cant (1983), in a model addressing the costs and benefits of foraging in Old World monkeys versus apes, define a food patch as a "localized aggregation of food items . . . separated from other such aggregations by regions of markedly lower food density" (Temerin & Cant 1983: 336). Baboons, macaques, vervets, and patas monkeys feed on a variety of food items, such as grass seeds, arthropod prey, and small herbaceous fruits, throughout the day (Agetsuma 1995b; Barton et al. 1993; Isbell 1998) while moving between food sites rather than feeding sequentially on single food types. Wrangham (2000) refers to this as a "feed as you go" strategy in describing bonobos in relation to common chimpanzees. Consequently, determining what is "markedly lower food density" and what, therefore, constitutes a patch is problematic (see Nakagawa 1989a). In order to assess the "patchiness" of foods available to savanna-dwelling primates, the entire combination of food items within the environment must be considered. The very fact that foods such as THV are not easily recognized as patches has caused researchers to conclude that these foods are less monopolizable than those that occur in easily identifiable patches (i.e., tree crowns) (Wrangham 1983). Recently, however, primatologists have questioned the ubiquity of THV and its hypothesized influence on aspects of primate socioecology—for example, group formation among great apes. Doran and McNeilage (1998) measured the availability of THV to western lowland gorillas in Central African Republic and concluded that this food source was actually clumped and patchily distributed.

An issue crucial to the study of primate behavioral ecology is the reliability of generalizing results across sites. Byrne and co-workers (1990) attempted to compare food availability for baboons across sites by using chemical assays of foods and strip plots to record food plant densities. These data provide information on food quality and abundance but not on food distribution at a scale relevant to a foraging individual (e.g., how far apart are food sources?). Similarly, Collins and McGrew (1988) used transect methods to compare the forests of chimpanzees studied at different sites in western Tanzania. These authors provided data on a much larger scale, specifically categorizing the different habitat types characterizing

Adult female patas monkey feeding. *(Photo by the author)*

three sites. Plotless methods (e.g., PCQ measures, distance to nearest neighbor plant species or feeding species) are often used to assess the distribution of feeding species and are easily comparable across sites. For example, the distribution of trees measured using plotless methods is reported in absolute amounts, such as the distance in meters (a standard measure) between feeding trees. These methods assume a random distribution of the organism studied however (Clarke 1986), and thus may not be appropriate for measuring the distribution of foods hypothesized to occur in clumps or patches (e.g., some fruiting trees). Assessing the distribution of food items within feeding tree crowns is more problematic, and measures are rarely comparable from one site to another. For example, Sterck and Steenbeek (1997) used estimated counts of food items as well as percentages of the different food types (e.g., fruits and flowers) within tree crowns to determine the abundance of foods available to langurs and macaques. The distribution of such foods within trees, however, was measured on a relative scale (Sterck & Steenbeek 1997). Comparisons of measures of the distribution of foods within trees will only be reliable using a relative scale when interobserver bias is eliminated (i.e., there is only one observer, or observers assess their reliability systematically). The distribution of foods on a small scale, however, like that of a single patch, may be most relevant to feeding competition at the level of an individual.

A primate's interpretation of the availability of any one food species or type may vary throughout the year as the relative availability of foods changes (*C. aethiops:* Harrison 1984; *Macaca* sp.: Agetsuma 1995b; *Gorilla* sp.: Remis 1997; Tutin & Fernandez 1993). Consequently, food availability should be monitored over the long term, minimally one annual cycle. Olupot (1998) found that mangabey (*Cercocebus albigena*) feeding behavior varied significantly over time depending on which food species were included in the diet, but the inclusion of different food types did not vary. Harrison (1984) found that secondarily preferred foods, such as leaves, gum, seeds, and fungi, were not eaten in proportion to their availability by vervets in Senegal but in relation to the availability of preferred foods, such as fruit or flowers. Stevenson (2004) suggests that the variation (>31%) not explained by more than 20 ecological variables examined in his study of fruit choice by woolly monkeys is likely because of the availability of alternate foods. These examples demonstrate the importance of distinguishing between food types rather than just between species, as well as considering the availability of alternative foods.

Food type, which may be used to indicate food quality [e.g., in Wrangham's (1980) model] may be as important as or more important than food distribution and abundance in influencing primate behavior. Sterck & Steenbeek (1997) concluded that the combination of reduced aggression among adult female Thomas langurs and the large dimensions of their feeding patches indicated that langur foods were abundant. This assumption did not hold for sympatric macaques in the same study, however. Neither patch size nor abundance but food type was the most important variable influencing aggression among adult female lion-tailed macaques (Sterck & Steenbeek 1997). If only food abundance and patch size had been considered for macaques, the authors would have failed to account for an important variable affecting the perception of food availability by female macaques—food type.

A common practice in the primate literature is to make broad generalizations with respect to the availability of certain food types. For example, the common assertion is that fruits are a rich source of energy for primates but are patchily distributed, whereas leaves are more densely and evenly distributed but are of lower dietary quality (Agetsuma 1995; Dunbar 1988; Oates 1987; Robinson & Ramirez 1982). Barton and co-workers (1993), however, point out that the traditional frugivory–folivory dichotomy does not take into account the fact that many fruits, such as those eaten by savanna baboons (and vervet and patas monkeys), are not of the type commonly alluded to, namely sugary berries or drupes. Even folivorous species are rarely exposed to a hyperabundance of foods. For example, mantled howling monkeys (*Alouatta palliata*) feed preferably on young leaves, which are not necessarily more abundant than other food items (Glander 1978; Milton 1980).

Underlying the importance of food type in regard to the assessment of food availability is the nutritional value of primate foods (see Fossey &

Harcourt 1977). Chapman (1985), in a study of vervets on St. Kitts, concluded that food abundance did not significantly influence food-site selection because vervets did not forage longer in areas of highest food plant density. Because vervets selectively fed in areas of secondary growth, Chapman (1985) suggested that food quality was the determining factor in food site selection. Using phytochemical data, Barton and co-workers (1993) concluded that similar factors influenced selection of foods by olive baboons in Laikipia, Kenya. These authors found no correlation between the abundance and the quality of baboon foods. Foods considered to be low quality (low in protein, high in fiber) were not more readily available than high-quality (high protein, low fiber) foods (Barton et al. 1993). These findings are in contrast with theories that generally assume that low-quality foods are ubiquitous.

Variation between plant food species and even individual variation within a single species of plant may influence primate behavior. For example, orangutans (*Pongo pygmaeus*) were found to be more efficient foragers in one *Ficus* species compared with a second species (Utami et al. 1997). The *Ficus* species with larger fruits was a more profitable food source for the orangutans because the foods were more efficiently processed (Utami et al. 1997). Glander (1978) has shown that even for a single tree species, individual variation in the production of secondary compounds may have important implications for primate behavior. Mantled howling monkeys in Costa Rica regularly fed on mature leaves of particular individuals of a certain tree species but bypassed other individuals. Even edible mature leaves, because of the presence of secondary compounds, may be widely distributed in that they are effectively less abundant per individual plant than if phytochemistry were not an important factor affecting primate food choice (Glander 1978). Chapman and co-workers (2004) found similar results in Uganda, where nutritional value of leaves, particularly protein, from plant species eaten by red colobus (*Piliocolobus tephrosceles*) and black and white colobus (*Colobus guereza*) varied between individual trees by as much as 20%. Clearly, the relationship between food type and availability is not always one that is simple and predictable.

## Clumped Resources and Within-Group Contest Competition

When primates' foods are monopolizable or usurpable (Isbell 1991; van Schaik 1989; Wrangham 1980), contest competition among females is expected to occur. Theoretically, such resources include those that are distributed in a manner that allows one individual to effectively keep others away. The resources may be of high nutritive value, abundant locally but scarce generally, or reflect some combination of such factors that induce one individual to usurp these foods from another individual. Although contests over foods have been found to only partially reflect the availability of foods contested (e.g., contests also serve to enforce social dominance that may or may not function primarily to allow preferred access to food resources; Johnson 1989; Radespiel et al. 1998; Silk 2003), models of female primate social behavior predict that contested foods will exhibit particular characteristics (Isbell 1991; van Schaik 1989; Sterck, Watts, & van Schaik 1997; Wrangham 1980).

Determining the monopolizability of a food has been, for the most part, a post hoc endeavor, and authors hold differing views as to which variables best predict this trait. Isbell and Young (2002) distinguish between monopolizability and usurpability. Usurpable foods are associated with agonistic interactions. Monopolizable foods are those for which an individual maintains possession of a food patch. They distinguish usurpability as a measurable behavioral consequence of the nature of a resource rather than monopolizability, because this condition requires no interaction between individuals: that is, monopolizability of a food resource does not indicate in any way that a resource will be contested but merely that it could be held by another individual (Isbell & Young 2002). For example, a single gum-feeding site may be monopolizable on the level of the individual primate but is not usurped because of the presence of multiple sites that enable all individuals within a group to feed simultaneously. The question then becomes, what makes a food worth usurping? What factors preclude or induce competition over foods?

In general, models of female primate social behavior hypothesize that spatially clumped and locally abundant resources elicit contest competition (Isbell 1991; van Schaik 1989; Sterck, Watts, & van Schaik 1997; Wrangham 1980). Evidence in support of this prediction has been found for macaques (*Macaca fascicularis*: Gore 1993; Sterck & Steenbeek 1997; *M. fuscata*: Ihobe 1989; *M. maurus*: Matsumura 1998, *M. mulatta*: Mathy & Isbell 2001), orangutans (Utami et al. 1997), baboons (*Papio* sp.: Barton 1993; Barton, Byrne, & Whiten 1996; Gore 1993), chimpanzees (Bygott 1979; Goodall 1986), mountain gorillas (*G. g. berengei*: Fossey & Harcourt 1977), patas monkeys (Struhsaker & Gartlan 1970), and vervets (Whitten 1983; Wrangham & Waterman 1981). For example, displacements among orangutans in Sumatra were observed only in and around large food trees, and subordinate individuals had restricted access to food sources compared with dominant individuals (Utami et al. 1997). A captive study that experimentally manipulated the distribution and abundance of foods to captive rhesus monkeys and Hamadryas baboons showed that competition within groups was most likely to occur when food was dispersed and scarce compared with "normal" and "rich" feeding conditions (Gore 1993). For patas monkeys in Cameroon, agonism rates varied seasonally with the availability of clumped resources (Struhsaker & Gartlan 1970). Agonistic encounters increased significantly during the dry season when fewer water sources were available and the number of monkeys at a single water source increased (Struhsaker & Gartlan 1970). In a comparison of two baboon species (*P. h. ursinus* and *P. h. anubis*), Barton and co-workers (1996) found that the proportion of clumped foods in the diet, such as seeds and flowers of *Acacia* and sedge corms, was an indicator of rates of supplanting between adult females during feeding. Female olive baboons in Laikipia, Kenya, spent more time feeding on clumped foods and contested foods significantly more often than female chacma baboons at Drakensberg, South Africa, who fed on dispersed foods (Barton, Byrne, & Whiten 1996). Clumped foods in this case were defined as those food sources that would permit a few group members but not all individuals to feed simultaneously and that were high in food density within clumps. In most cases, clumped foods are thought to be somewhat isolated in space but abundant within a single patch, but an operational definition is not always supplied. A recent experiment by Mathy and Isbell (2001) suggests that food size, quality, and depletion time are variables of greater importance than spatial distribution in eliciting contest competition in primates.

A combination of the variables discussed in this section should be considered to thoroughly understand the relationship between a primate's behavior and its ecology. Variables such as primate nutritional requirements; physical and social limitations to primate feeding behavior; phylogenetic constraints on diet, social system, and behavior; plant phytochemistry; food processing costs; satiation; diet variability; and food distribution and abundance interact to produce the observed patterns of behavior that are often simplified into a single or small number of categories.

# 3

# Foods Available to Vervet and Patas Monkeys

To assess the influence of food availability on vervet and patas monkey feeding behavior, contest competition, and dominance relationships, I collected data on the abundance and distribution of their foods. Food availability is invoked by primatologists to explain primate behavior such as feeding and foraging, ranging, grouping behavior, and social interaction; therefore, to better understand the relationship between food availability and such behaviors, researchers should establish which methods are most accurate under particular conditions (e.g., in tropical forests versus savanna habitats). Hence, I appraise the utility of different methods in estimating food availability for vervet and patas monkeys in the habitats they use on Segera. Finally, I assess whether differences in food availability exist within the environment used by vervets (whistling-thorn and fever-tree habitats) and between the whistling-thorn woodland habitats used by vervet and patas monkeys.

## HYPOTHESES AND PREDICTIONS

The hypotheses that I tested were related to potential differences in the abundance and distribution of foods in the various habitats within the home ranges of the study groups. In addition to testing the hypothesis that predicted that foods would be more clumped in the riverine compared with the whistling-thorn woodland for vervets, I also tested the null hypothesis that the distribution and abundance of whistling-thorn foods did not differ significantly between the vervets' and patas monkeys' home ranges. Even if both species at this site use whistling-thorn woodland that is similar, the availability of foods in different areas of this habitat may differ in ways that significantly vary their behavior. For example, the diversity of feeding plant species or the relative number of food plants may contribute to differences in feeding behavior (Budnitz 1978).

Whistling-thorn *Acacia* trees support several obligate ant species (*Crematogaster* sp., *Tetraponera* sp.) that defend particular parts of the trees, including food items important to primates (Madden & Young 1992; Young, Stubblefield, & Isbell 1997). Ant species attack ungulate browsers like giraffes (Dagg & Foster 1976; Madden & Young 1992) and other predators such as primates, and different ant species vary in the degree to which they aggressively defend plant parts (Young, Stubblefield, & Isbell 1997). The presence of obligate *Acacia* ant species is therefore likely to reduce the potential harvest (after Sterck & Steenbeek 1997) available to monkeys. Thus, whereas individual counts of food items on trees may represent the *potential* harvest available to primates, the influence of obligate ant species must be considered to determine the *actual* harvest available as perceived by the monkeys. I predicted that the ant species resident on whistling-thorn feeding trees would affect vervet and patas monkey behavior, and therefore I collected data on the distribution and abundance of the ant species resident on whistling-thorn *Acacia* and the relative aggression of the different species.

I used a number of measures of food distribution (PCQ, quadrat, strip plot) and abundance (counts of food items, tree height, and crown dimensions) to record the availability of foods to monkeys. I used plot samples and plotless samples to measure the availability of whistling-thorn trees. Counts of food items on whistling-thorn trees were used to measure the abundance of these foods (swollen thorns, gum, flowers, new growth/leaf shoots, seed pods). I compared traditional measures of food availability, such as tree height and crown diameter, with counts of food items to assess their usefulness in predicting food abundance. Additionally, by reviewing available data from the literature, I examined food quality as a variable potentially affecting how primates would perceive the foods available to them.

## DEFINING A FOOD PATCH

Theories of contest competition among female primates base their predictions on the monopolizability of food patches, but it is difficult to define what makes a food patch monopolizable (or usurpable) except by *post hoc* analyses of feeding competition. I assumed that patas and vervet monkeys could monopolize an area in which they could move and eat simultaneously. I used White and Wrangham's (1988) definition of a patch (i.e., an area in which an animal can continuously eat and move), which allows the traditional definition of a patch as a single tree crown to be extended to include multiple crowns (see Chapman 1988). In regard to whistling-thorn trees, one to four crowns most often constituted a patch (Pruetz 1999). If tree crowns were adjacent, so that a subject could move and forage between crowns while feeding, they were considered a single patch.

Based on preliminary observations at the Segera site, the dimensions of whistling-thorn patches were hypothesized to be monopolizable by individuals, whereas other food species such as fever tree *Acacia* and *Scutia myrtina* shrubs would preclude monopolization by one individual because they were too large. I used the concept of "personal space," which has been defined for female baboons as an area in which few individuals are allowed (Altmann 1980), to designate an area from which other individuals could be excluded (i.e., a "feeding space"). I assumed that such an area would be defended from other individuals. To estimate the number of monkeys that could feed simultaneously at a feeding site, I calculated a feeding space based on theories of the size of an individual's personal space in baboons (Altmann 1980; Ron, Henzi, & Motro 1996). Among female chacma baboons, individuals generally kept distances of 2–5 m from nearest neighbors while feeding (Ron, Henzi, & Motro 1996). Altmann (1980) deemed 2 m as a distance designating "personal space" in female yellow baboons (*P. h. cynocephalus*). Assuming 2 m is an accurate estimation of the personal space for baboons, and because baboon body length is approximately one-half this distance (Altmann 1980), I calculated personal space for patas and vervet monkeys by doubling individuals' body lengths. Accordingly, adult female vervets would have an "approach radius" of approximately 1 m (head and body length, 40–61 cm: Haltenorth & Diller 1977). Mathy and Isbell (2001) also deemed 1 m to be the distance at which rhesus monkeys were able to exclude individuals from a desired food source. Body length for this species is 44–53 cm (Groves 2001), giving rhesus monkeys an approach radius of approximately 1 m according to my calculations and coinciding with the distance that Mathy and Isbell (2001) found to result in monopolization of a food source by an individual. Feeding spaces for adult female vervets (based on the formula for the volume of a sphere: $4\pi r^3/3$) would equal 4.2 m³. Adult female patas monkeys would have an approach radius of approximately 1.1 m (head and body length, 50–60 cm: Haltenorth & Diller 1977). Using this calculation, the feeding space for adult female patas monkeys would be approximately 5.6 m³. I predicted that an intrusion into this space would incite feeding interference. This space was larger for both monkey species than the volume of most whistling-thorn patches, which averaged 3.6 m³ for vervets and 4.1 m³ for patas monkeys (Pruetz 1999).

When formulating the definition of a whistling-thorn patch, I hypothesized that obligate ant species influenced primate feeding and ranging behavior. I predicted that obligate ants resident on whistling-thorn trees with contiguous crowns would be affected by the disturbance of a predator, such as a monkey, feeding anywhere within this patch. The monkeys were expected to perceive overlapping crowns as areas in which foraging is limited once the patch is disturbed. The costs of usurping food patches such as these, in which the foraging window is small because of ant attack, were expected to be high. In other words, I predicted that it would be less costly for an individual to move to an undisturbed feeding site rather than usurp an already disturbed site where foraging time is shortened by ant defense. Taking into account relevant behavioral information (e.g., how large an area an individual could defend) and ecological information (e.g., the influence of ant protection on feeding behavior) enabled me to quantify food availability at a more accurate level or scale.

## MEASURING FOOD AVAILABILITY

To assure accurate estimation of the foods available to vervet and patas monkeys, as well as the reliability of these measures, I used many different types of data collection methods to assess the availability of food resources used by monkeys (Tables 3–1 and 3–2).

**Table 3–1** Measures of Food Availability: Whistling-Thorn *Acacia*

| Availability Type | Method(s) Used |
|---|---|
| Tree density/abundance | Permanent quadrats ($N = 16$) monitored monthly |
| | Monthly quadrats along transects ($N = 58$) |
| | Variation on PCQ ($N = 894$ data points) |
| Tree distribution | Permanent quadrats ($N > 500$ trees) |
| | Variation on PCQ ($N = 894$ data points) |
| Food item abundance | On trees along transects ($N = 258$ trees) |
| | On trees in permanent quadrats, presence/absence (mean 375 trees per month) |
| | On feeding trees from bout sampling ($N = 570$ trees) |
| | On trees sampled for ant aggression, presence/absence of foods ($N = 554$ trees) |
| Properties of foods | Swollen thorn contents ($N = 114$ thorns) |
| | Gum patch dimensions ($N = 12$ trees, 97 patches) |
| *Acacia* ant aggression | Ant aggression bouts ($N = 554$ trees) |
| *Acacia* ant species density | On trees in permanent quadrats ($N = 542$); sampled for food abundance ($N = 258$ trees), for swollen thorn contents ($N = 114$) |
| | Ant species on trees fed in during continuous focal animal sampling (from 435 samples) and feeding bouts ($N = 570$ trees) |
| | Ant species on trees in two 100-tree transects in patas monkeys' range ($N = 200$ trees) |

**Table 3–2** Measures of Other Important Foods' Availability

| | |
|---|---|
| Fever tree *Acacia*—distribution and abundance | Variation on PCQ, along transects ($N = 80$ data points) |
| Fever tree *Acacia*—gum abundance | On individual trees ($N = 9$ trees, 87 gum patches) |
| THV (including *Commelina, Hibiscus, Solanum, Ipomoea*)—distribution and abundance | 1 m² quadrats in permanent quadrats ($N = 76$) <br> 1 m² quadrats along transects, monthly ($N = 132$) |
| *Scutia myrtina*—distribution and abundance | Variation on PCQ along transects ($N = 60$ data points) <br> Abundance of fruits within shrubs ($N = 16$) |
| *Brachiaria brizantha*—abundance and distribution | 1 m² quadrats along transects ($N = 73$) |

## Large-Scale Food Availability

*Methodology*   Several different measures were used to record the density/ abundance and distribution of whistling-thorn trees within the home ranges of the study groups. Plot sampling included monitoring sixteen different "permanent" quadrats of 10 m² each month. The goal in establishing these quadrats was to sample approximately two hundred individual trees within the home ranges of each species in areas easily accessible by vehicle so as to maximize collection of data on a large number of individual trees per month. Permanent quadrats were used to monitor the seasonal availability of primate foods. The number of quadrats sampled in each species' range depended on the density of trees in the different quadrats. Approximately two hundred whistling-thorn trees in four permanent quadrats—three of dimensions 10 m × 10 m and one of dimensions 5 m × 6 m—were monitored monthly within the area of home range overlap between the Pond and Glade groups of vervets. Approximately two hundred whistling-thorn trees in eleven permanent quadrats of 10 m² were monitored monthly within the patas monkeys' home range. A mean of 375 trees (range 250–450) was sampled each month for 14 months from October 1993–December 1994, excluding July 1994. Additional data were collected on some of these trees in September 1993 and January–March 1995.

I used a variation on the PCQ method, a plotless sampling method, to measure tree distribution every 20 m along transects ($N = 48$ points). The location of these transects varied each month, according to the monkeys' ranging behavior. This same plotless method was used by Whitten (1988) to assess the distribution of trees within the home ranges of vervets in Samburu, Kenya. Such methods assume the random distribution of the organism studied (Clarke 1986), and it was hypothesized that whistling-thorn trees were randomly distributed. Transects were chosen using a stratified random sampling method for patas monkeys in which map coordinates spaced 50 m apart were randomly chosen from areas used by the monkeys during the previous month. Because vervets typically used all of their home range monthly, transects oriented east–west within their home range were sampled at different points each month. These transects were located every 75 m along a north–south trajectory following the Mutara River (i.e., transects bisected the vervets' home range, including riverine as well as woodland). The space around the designated point was divided into four equal and infinite sections in which the distance in meters to the nearest tree was measured (i.e., quadrants). My variation on the PCQ method entailed measuring the distance from a transect point to the nearest tree in each of five different height categories, because I predicted that trees of various heights would also vary in the amount of food they produced. I classified height categories by 0.5 m intervals up to 2 m. I classified all trees taller than 2 m into one category, because of the scarcity of many tall trees (see also Dagg & Foster 1976). Each sampled point yielded a total of 20 data points (five different height categories × four quadrants).

A third measure of the abundance and distribution of whistling-thorn trees was plot sampling along the same transects as described earlier (i.e., strip plots). At the same points every 20 m along transects where the PCQ samples were taken, a 10 m² quadrat was used to quantify the number of *Acacia* trees present and their heights. The PCQ and strip plot methods were directly comparable because they measured tree densities at precisely the same sampling points.

*Results* The PCQ and strip plot methods differed significantly in the density of trees measured per hectare (patas monkeys—$\chi^2$ = 892.2; degrees of freedom, df = 1; $p < .01$; vervets—$\chi^2$ = 440.9, df = 1, $p < .01$). The proportions of the densities of trees of different heights were similar for trees within the patas monkeys' range when using the strip plot and PCQ measures. The methods differed significantly, however, regarding the frequencies of trees within the different height classes (df = 1, $p < .01$), with the exception of trees within the less than 0.50 m category in the vervets' home range (Figure 3–1).

Because data collected on tree densities from strip plots corresponded most closely with earlier studies of whistling-thorn tree densities on Segera (Dagg & Foster 1976—1440 trees/ha; Young, Stubblefield, & Isbell 1997—1335 trees/ha), these were taken to be the most reliable. Based on strip plot data, a mean of 2477 trees per hectare characterized the vervets' home range, compared with a mean of 1411 trees per hectare in the patas monkeys' home range. These densities differed significantly (independent *t*-test, $t$ = 2.424, df = 50, $p$ = .019). Compared with findings of whistling-thorn tree density from other studies, the home range of the vervets is an area of dense woodland. Whistling-thorn woodland, such as that used by vervets

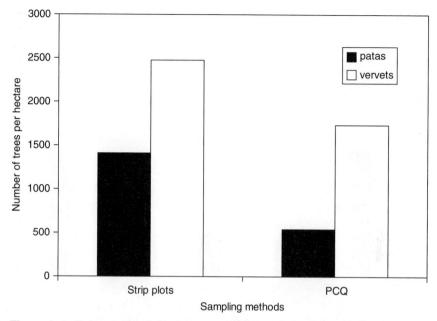

**Figure 3–1** Estimates of whistling-thorn trees per hectare using different ecological sampling methods.

bordering gallery forest, is characterized by shorter trees (Taiti 1992) that rarely exceed 1.5 m in height.

The whistling-thorn woodland profile differed somewhat when comparing the home ranges of the two species. Trees in the patas monkeys' home range were, on average, taller than those within the vervets' home range (Figure 3–2). The most common height class of trees within both species' home range included those less than 0.5 m in height, but for patas monkeys trees greater than 2 m in height were equally common. Trees greater than 2.0 m in height were least common for vervets. Foraging vervets encounter a greater number of trees and foods than patas monkeys in general, and most of these trees were short (Figure 3–3). Although taller trees provide more food items than shorter ones, the availability of ant-defended foods is independent of tree size (see Results). Monkeys feed on only a few items per tree of any size (see Chapter 4). Therefore, the availability of these foods was greater for vervets than for patas monkeys based on the overall density of trees.

In addition to being an abundant resource, based on estimated densities, whistling-thorn is also a scattered resource (Isbell et al. 1998b and this study), not occurring in large discreet patches or stands such as fever tree *Acacia*. The spatial distribution of whistling-thorn trees within the monkeys' home ranges did not differ significantly from random (patas monkeys' home range—$\chi^2 = 11.546$, df $= 13$, $p > .50$; vervets' home range—$\chi^2 = 22.11$,

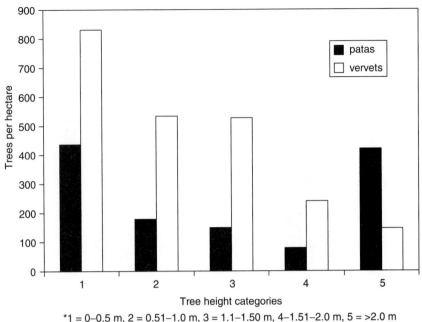

*1 = 0–0.5 m, 2 = 0.51–1.0 m, 3 = 1.1–1.50 m, 4–1.51–2.0 m, 5 = >2.0 m

**Figure 3–2** Frequency of whistling-thorn trees in different height classes (1 = 0–0.5 m, 2 = 0.51–1.0 m, 3 = 1.1–1.50 m, 4 = 1.51–2.0 m, 5 = > 2.0 m) between woodland habitats used by patas monkeys and vervets.

df $= 29, p > .80$) based on the divergence of the variation from the mean (Madrigal 1998). The spatial distribution of individual whistling-thorn trees, however, did differ significantly for patas and vervets based on data from PCQ measures ($t = 2.040$, df $= 24.9, p = .05$).

The PCQ method provides a more accurate measure of patterns of spacing between trees than do estimates of density. Cottam and Curtis (1956) compared a number of methods used to measure the distribution of individuals in space with absolute counts in three different forests and in an artificial population to determine the most accurate method. The PCQ method exhibited the lowest variation and required the least number of samples to yield a standard error of less than 5% of the mean. A range of 26–38 samples proved accurate using the PCQ method, whereas other methods required 50–71 samples (random pairs method), 59–102 samples (nearest neighbor), and 114–146 samples (closest individual). Additionally, calculating the mean distances between subjects for methods other than PCQ required using a correction factor (Cottam & Curtis 1956).

Trees within the patas monkeys' home range were more widely dispersed than within the vervets' home range based on PCQ measures. Average distance between individual whistling-thorn trees for vervets

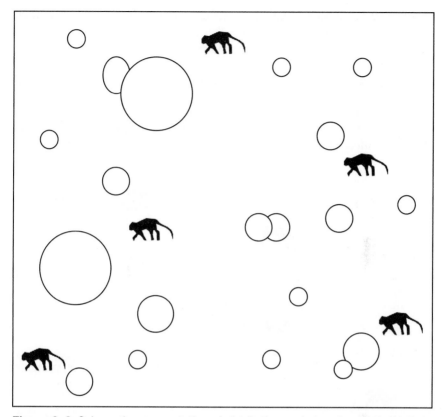

**Figure 3–3** Schematic representation of distribution and abundance of whistling-thorn *Acacias* of different sizes (circle size corresponds to tree crown size of different height categories; largest trees are characterized by crown diameter of ~2.5 m) available to foraging vervets on Segera.

was 2.4 m, compared with 4.3 m for patas monkeys. These results correspond with Isbell and co-workers' (1998b) behavioral measures of food distribution for these same study groups during this period. Vervets in whistling-thorn did not travel as far between feeding sites as patas monkeys (Isbell et al. 1998b), indicating that resources for vervets were not as widely distributed.

Within the vervets' range, I estimated the width of the riverine habitat to the nearest half-meter ($N = 16$ points) along the same transects that I used to measure the distribution of whistling-thorn. The riverine habitat averaged 61 m in width within the study groups' home range ($N = 16$ transect points; range 27–105 m). The Mutara is a very narrow river, averaging about 10 m in width. The riverine area totals about 112,800 m², or 28% of the vervets' 40 ha home range. The PCQ method was used to record data on the distribution of fever trees ($N = 20$) and *Scutia myrtina* ($N = 17$)

shrubs in the riverine habitat within the vervets' home range. PCQ methods showed that *S. myrtina* shrubs were distributed on average every 18 m at a density of 30 shrubs per hectare. The mean height of shrubs sampled ($N = 9$) was 2.5 m, and mean maximum crown width was 3–9 m ($N = 5$). Fever trees were distributed every 13 m on average here, based on data from PCQ methods ($N = 80$), at a density of 56.7 trees per hectare. Mean height of fever trees sampled was 5 m ($N = 78$), and average crown width was 5 m ($N = 67$). The distribution and abundance of major tree and shrub food resources therefore differed between the two habitats used by vervets (Figure 3–4).

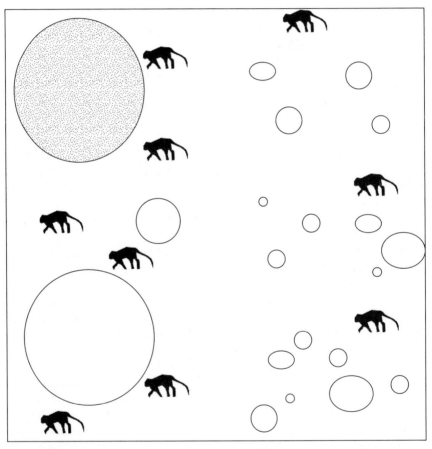

Riverine habitat            Whistling-thorn woodland habitat

**Figure 3–4** Schematic representation of major food tree and shrub species within riverine habitat versus whistling-thorn woodland used by vervets on Segera. Darkly shaded spheres represent *Scutia myrtina* shrubs; lightly shaded (larger) spheres represent fever trees (with an average crown diameter of 5.0 m); open spheres represent whistling-thorn *Acacia*.

## Small-Scale Food Availability

*Methodology*   To estimate food availability on a smaller scale, I collected data on different aspects of individual food species. I recorded data monthly on the abundance of food items in whistling-thorn trees from June 1994 to December 1994. Trees were sampled every 20 m along the same transects described earlier and trees within the five different height categories were sampled systematically. Random trees were sampled at every 20 m along transects, and swollen thorns in various stages of development (mature, immature) were collected ($N = 114$) so that data could be recorded on size and contents. Approximately two trees per height class (0–0.5, 0.5–1.0, 1.0–1.5, 1.5–2.0, and >2.0 m) in monkeys' home ranges were sampled per month ($N = 258$). Data collected per tree included (1) height of tree, (2) crown height and depth, (3) two perpendicular (one maximum) crown diameter measures, (4) obligate ant species resident, and (5) the number of food items available to primates, such as swollen thorns, new growth (leaf shoots and new swollen thorns), gum patches (i.e., contiguous areas of gum), seed pods, and flowers. Tree heights and crown dimensions were measured using a metered stick.

*Results*   I first determined the productivity of whistling-thorn trees within the monkeys' home ranges by averaging the abundance of food items for all trees regardless of height. The availability of swollen thorns, seeds, new growth, and flowers did not differ significantly between the species' home ranges when tree height was controlled in this way (Table 3–3). Availability of whistling-thorn gum within the patas monkeys' range was significantly greater than in the vervets' range even when trees of similar heights were compared. The most abundant food item available to both patas monkeys and vervets was swollen thorns, and gum was the least abundant food item for both. Within the respective height classes, trees in the monkey species' home ranges did not vary in dimensions, such as crown width, depth, or volume ($\chi^2$ tests: all ns, $p > .05$).

**Table 3–3** Abundance of Whistling-Thorn *Acacia* Food Items (Mean Number Per Tree)

| Food Item | Vervets Home Range | Patas Monkeys Home Range | Independent t-Test 2-Tailed |
|---|---|---|---|
| All food items | 154.1 | 139.7 | $p = 0.333$ |
| Swollen thorns—mature + immature | 120.1 | 113.0 | $p = 0.855$ |
| Seeds—mature + immature | 18.2 | 14.9 | $p = 0.660$ |
| New growth | 8.7 | 6.5 | $p = 0.218$ |
| Flowers | 8.6 | 4.3 | $p = 0.326$ |
| Gum | 0.4 | 2.1 | $p = 0.001$ |

**Table 3–4** Pearson Correlation Coefficients Matrix: Whistling-Thorn Tree Dimensions

| Dimension | Crown Diameter | Crown Volume | Tree Height | Crown Depth |
|---|---|---|---|---|
| Crown diameter | | | | |
| Crown volume | 0.813 | | | |
| Tree height | 0.758 | 0.486 | | |
| Crown depth | 0.710 | 0.425 | 0.929 | |

Analyses were conducted to assess the degree to which different characteristics of the trees that were measured correlated with one another (Table 3–4) as well as the value of different characteristics such as crown dimensions and tree height, in predicting the availability of different food items (see Table 3–4). Characteristics of individual whistling-thorn trees, such as height, crown volume, crown diameter, and crown depth, correlated positively with one another to varying degrees (see Table 3–4). Crown diameter was highly and positively correlated with all other dimensions measured. Tree height and crown depth were highly and positively correlated with one another. These variables, as well as crown volume—using the formula for volume of a hemisphere $[(4/3)\pi(D/2)^3/2]$, where $D$ = crown diameter—were then regressed against the availability of all foods and individual food items (Table 3–5). These results are especially important as they are based on absolute or known food availability as derived from counts of food items.

Food item availability varied with different aspects of tree size. The abundance of food items on whistling-thorn trees correlated with tree height, maximum crown width, crown volume, and crown depth (see Table 3–5). Maximum crown width was the best indicator of food item abundance. Crown width correlated highly and positively with ant-defended food

**Table 3–5** Pearson Correlation Coefficients Matrix: Whistling-Thorn Size and Food Item Availability

| | Tree Height | Crown Width | Crown Depth | Crown Volume |
|---|---|---|---|---|
| All foods | 0.602 | 0.732* | 0.536 | 0.577 |
| Ant-defended foods | 0.608 | 0.680* | 0.554 | 0.464 |
| Non-defended foods | 0.228 | 0.405 | 0.174 | 0.476* |
| Swollen thorns | 0.611 | 0.676* | 0.553 | 0.461 |
| New growth | 0.270 | 0.382* | 0.286 | 0.267 |
| Flowers | 0.179 | 0.271 | 0.178 | 0.358* |
| Gums | 0.346* | 0.222 | 0.255 | 0.155 |
| Seeds | 0.151 | 0.325 | 0.082 | 0.366* |

*Best indicator of abundance.

item abundance (i.e., swollen thorns, new growth) and when all food items were considered. Tree height was the best predictor of the availability of gum to monkeys (see also Isbell 1998). Crown volume was the best predictor of non-defended foods (i.e., flowers, seed pods). Tree height was originally thought to be an important indicator of food availability, but crown width and, subsequently, crown volume would have been stronger measures of the abundance of foods, with the exception of gums. These results illustrate the importance of evaluating several measures to determine their accuracy in quantifying the production of foods important to primates.

The availability of food items within trees of different heights is shown in Table 3–6. Mature swollen thorns were the most abundant food in trees of all heights within both species' home ranges. Gum was more abundant within the home range of the patas monkey study group even when height of the tree was controlled. Trees of all heights contained more gum in the patas monkeys' home range than in the vervets' home ranges. Burning within the home range of patas but not vervets may have contributed to the availability of gum, because gum flow is stimulated by damage to the tree trunk and branches (Coe & Beentje 1991). Additionally, tall trees in the patas monkeys' range contained more gum than shorter trees. Trees greater than 2 m in height averaged eight patches per tree. Trees of such heights are fairly widely distributed, and because tall trees contained a large number of patches compared with other trees, gum is a relatively clumped whistling-thorn food. Thus, gums can be termed "patchy" but only in relation to other whistling-thorn foods (Figure 3–5). Because gum was an important food for patas monkeys (Isbell 1998), the dimensions of gum patches (i.e., contiguous areas of gum) were measured. During August 1993, trees from which monkeys ate gum ($N = 9$), in addition to several randomly selected trees in the same vicinity ($N = 3$), were sampled regarding their gum production. The height of these trees and data on

**Table 3–6** Whistling-Thorn Foods in Trees of Varying Heights: Estimated Number per Patch

| | Patas Monkeys (Height Classes) | | | | | Vervets (Height Classes) | | | | |
|---|---|---|---|---|---|---|---|---|---|---|
| | 1 | 2 | 3 | 4 | 5 | 1 | 2 | 3 | 4 | 5 |
| Swollen thorns | 13 | 66 | 147 | 197 | 303 | 19 | 69 | 138 | 189 | 317 |
| New leaves | 3 | 5 | 9 | 10 | 11 | 3 | 7 | 13 | 12 | 18 |
| Seeds | 0 | 1 | 12 | 59 | 16 | 0 | 12 | 42 | 29 | 21 |
| Flowers | 0 | <1 | 1 | 21 | 0 | 0 | 4 | 7 | 37 | 7 |
| Gum | 0 | 1 | 1 | 2 | 9 | 0 | <1 | <1 | 1 | 3 |
| All foods | 16 | 63 | 162 | 290 | 308 | 20 | 89 | 199 | 265 | 374 |

Height classes: 1 (0–0.5 m), 2 (0.51–1 m), 3 (1.01–1.5 m), 4 (1.51–2 m), 5 (>2.0 m).

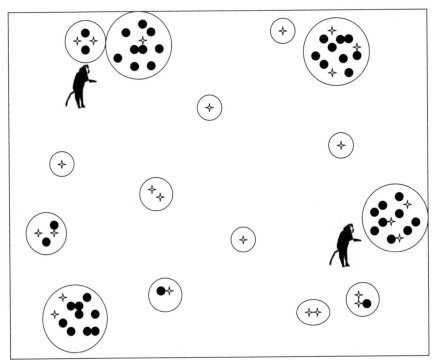

**Figure 3–5** Distribution and abundance of whistling-thorn gum and thorns available to foraging patas monkeys. Open circle size corresponds to tree crown size of different height categories. Largest trees are characterized by crown diameter of ~2.5 m. Filled circles represent average number of gum patches available per tree in the various height categories. Stars represent average number of swollen thorns available per tree in the various height categories after taking into consideration obligate *Acacia* ant defense of foods.

gum patches, including length and width of patches and a description of the consistency of the gum (flowing, soft, hard) were collected.

Although food items other than gums did not differ significantly in abundance for patas monkeys and vervets when height of the tree was controlled, food item availability did differ because the woodland profile (i.e., density and height of trees) differed between the two species' home ranges. An estimate of the number of whistling-thorn foods available to patas monkeys and vervets (based on mean values of foods available in trees of certain heights and data on the distribution of these trees within the primates' home ranges) shows that vervets have approximately 36% more foods available per square meter than patas monkeys. This estimate, however, does not take into account the effect of ant defense on food availability.

## Swollen Thorns and Obligate Ants

Because animal prey within the swollen thorns was important to patas monkeys (Chism & Wood 1994; Isbell 1998) and vervets (Isbell et al. 1998b), data were collected on the food content of individual swollen thorns. One to two thorns from individual *Acacia* trees or patches were sampled every 20 m along transects within both the species' home ranges ($N = 114$). Sixty-two thorns were collected during wet season months, and fifty-two thorns were collected during dry season months. Data collected on the tree/patch from which the swollen thorn was taken included (1) tree height, (2) two perpendicular (one maximum) crown width measures, (3) crown depth, (4) obligate ant species resident, and (5) height. Data collected on each swollen thorn included (1) growth stage of the thorn (immature, mature) after Young and co-workers (1997) and (2) the contents of the thorn, including the number of adult ants, alates, larvae, and eggs. Obligate ant species inhabiting the thorn were identified after Young and co-workers (1997), based on body size and color of the different body parts (head, thorax, abdomen).

Immature thorns comprised 57% of the total swollen thorn sample, whereas 43% of the thorns collected were classified as mature (after Young, Stubblefield, & Isbell 1997). Mature and immature swollen thorns did not differ significantly in the amount of ant foods (number of adults, larvae, alates, and eggs) contained within the thorns, although a trend for immature thorns to contain more ant foods than mature thorns was apparent ($t = -1.785$, df $= 19.4$, $p = .090$). The mean number of ant foods available to monkeys was 66.7 per thorn, with adults being most common (mean number of adults per thorn $= 33$, see Pruetz 1999).

Aspects of the obligate *Acacia* ants' natural history, such as residency patterns, were also studied to better understand these species' interactions with primates. Information on ant distribution was collected on (1) trees sampled using a variation on the PCQ method along transects, (2) trees sampled monthly along transects for data on the abundance of whistling-thorn food items, (3) trees sampled along two 100-tree transects (approximately 400 m in length) within the range of the patas monkeys during the wet season in two different areas, and (4) trees marked in the permanent quadrats. I used all trees in quadrats sampled during June 1994 to represent patterns of ant species residency. This time period was chosen merely as a point approximately halfway through the study when most permanently marked trees were surveyed within a single month. Data collected included information on ant species resident in a patch and characteristics of that patch, such as height and crown dimensions. The distribution of the various ant species differed. The most common genus was *Cremato-gaster* (Figure 3–6). The distribution of the different ant species inhabiting

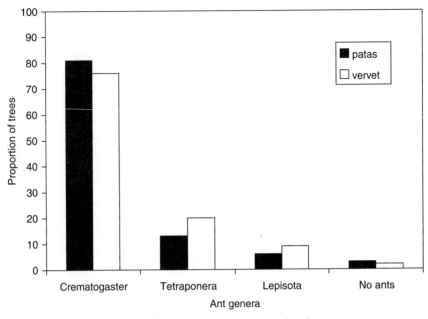

**Figure 3–6** Ant species distribution on whistling-thorn *Acacia*.

trees sampled primarily for food item abundance was similar to the data on ant residency collected along transects within the patas monkey home range. Trees within the vervet and patas monkeys' home ranges differed in the distribution of the various species of obligate *Acacia* ants. Vervets' trees had more nonaggressive *Tetraponera* species as residents ($\chi^2 = 20.45$, $p < .01$, df = 1) than did patas monkeys' trees (see also Pruetz 1999).

The obligate ant species resident on whistling-thorn trees were hypothesized to effectively restructure the pattern of availability of the monkeys' major food resource. A seemingly clumped resource becomes dispersed if primates are able to feed only on a few items per patch because of ant protection. Therefore, data were collected on the aggressive behavior of the different ant species inhabiting these whistling-thorn trees (after Madden & Young 1992) to compare with monkeys' interactions with the ants and trees. A minimum of ten trees per month in each primate species' home range was sampled, with one tree and/or patch being sampled every 20 m along transects ($N = 554$). Approximately one out of three trees was sampled at least twice, noting the amount of time elapsed between disturbances and distance (in centimeters) between the food items disturbed, so as to note the effects of previous disturbances on ant activity within a tree or patch. Different types of foods as well as nonfood items (i.e., bare branches or trunk) were disturbed at different heights and locations within the tree crown by tapping a dead branch or twig directly on the item (after

Madden & Young 1992). Data collected included (1) time of day, (2) height of tree, (3) tree crown height, and (4) two opposite (one maximum) crown diameter measures, (5) ant species resident, (6) number of seconds since disturbance of food item (from 5 to 70 s), (7) food item type disturbed, and (8) number of ants attacking.

*Crematogaster* species showed higher levels of aggression than either *Tetraponera* or *Lepisota* species, with an average of fifteen ants attacking every 10 s after disturbance compared with less than 2.5 ants every 10 s, respectively (analysis of variance, $f = 11.948$, df $= 3$, $p < .001$). Data were categorized and analyzed according to time of day to assess whether ants were more active at certain times. Four different time blocks were analyzed: 0700–0959 hours, 1000–1259 hours, 1300–1559 hours, and 1600–1800 hours. Analyses of variance tests showed no significant differences according to time of day ($f = 1.933$, df $= 3$, $p = .114$), concurring with a report by Madden and Young (1992). Eighty-two patches were disturbed more than once to examine the effect of previous disturbance on ant activity within a patch. The amount of time lapsing between disturbances was recorded as well as the distance between disturbance sites. An analysis of variance (ANOVA) test showed that ants from predisturbed trees ($f = 0.404$, df $= 1$, $p = .525$) or patches ($f = 0$, df $= 1$, $p = .997$) were not more aggressive than ants from trees not previously disturbed.

### Individual Foods in Riverine Habitat

Additional data on food species within the riverine habitat were also collected. This was done using a smaller spatial scale. The distribution of fruits of *S. myrtina* within shrub crowns was calculated by quantifying the number of ripe and unripe fruits (berries) available to vervets within a $0.25 \text{ m}^3$ area at different locations in the crown (inner, outer, upper, lower) and at different heights ($N = 21$). The heights at which $0.25 \text{ m}^3$ samples were taken ranged from 0.4 to 1.75 m. The distribution of fruits within *S. myrtina* shrubs ($N = 16$) sampled did not differ significantly from a random distribution for either ripe ($\chi^2 = 15.835$, df $= 20$, $p > .70$) or unripe ($\chi^2 = 15.77$, df $= 20$, $p > .70$) fruits. A mean crown volume of 31 $\text{m}^3$; characterized shrubs sampled. The average number of ripe fruits available to vervets within the sampled area was approximately 6.1, whereas unripe fruits occurred at a mean density of 39 per $0.25 \text{ m}^3$. Average fruit production was subsequently estimated as 756 ripe fruits and 4,836 unripe fruits (using the formula for the volume of a sphere) per individual shrub.

I took advantage of fever trees recently felled by elephants ($N = 9$) to quantify gum patches and patch size dimensions because this was not possible otherwise. Average crown width of these trees was 7.7 m, and mean tree height was 11.5 m. Trees that had already reacted to elephant damage (with visibly increased gum production at site of damage) were not included. The nine trees sampled contained a total of eighty-seven gum patches. Data collected on gum patches included length and width. Seventy-five gum patches were measured within these nine trees (mean 8.3 gum patches per tree). Ten patches were not included in the analyses because they were caused by recent elephant damage (and thus not representative of gum available to vervets in standing trees).

## Nutritional Value of Foods

I compiled data on the nutritional value of foods eaten by patas and vervet monkeys using published sources from studies of vervets and baboons in East Africa (Table 3–7). Nonfood plant species or those that did not occur in the home ranges of monkeys on Segera but were in the same genera as food species were included in some cases for comparison. Foods highest in protein included seedpods of *Acacia seyal*, flowers of *A. mellifera* and *A. tortillis*, and leaves of *Lycium europeum* and *Pennisetum* species. Values of foods for ash content, lipids, and other variables reported were less variable than those for protein values (see Table 3–7).

*Herbaceous-Level Food Availability*  Small quadrats of roughly 1 m² are often used to quantify herbaceous-level vegetation (Clarke 1986). Researchers at Chololo, Laikipia, used such quadrats to measure herbaceous foods important to savanna-dwelling olive baboons in Kenya (Barton & Whiten 1993; Marsh 1993). I used this method to record the presence or absence of food items besides *Acacia*, as well as the number of individuals of these species and number of food items (e.g., five plants of *Hibiscus flavifolius* with eight fruits) in plots nested within each of the sixteen 10 m² permanent quadrats each month for six months ($N = 76$). Additionally, during each of three dry season and three wet season months, 1 m² plots ($N = 132$) were sampled to quantify herbaceous-level food species and food items in both species' home range along transects. Plots were located every 20 m along transects within the patas monkeys' home ranges in areas that were used by them during that month's sampling period and along east–west transects within the Pond vervet group's home range (every 75 m along a north–south trajectory).

   An additional food important to vervets within the riverine habitat was *Brachiaria brizantha* grass. *Brachiaria* grass grows alongside river edges during the wet season as extensive mats and forms a mostly pure population (rating 5 on Braun-Blanquet Sociability scale: Clarke 1986). Data on the area of patches (contiguous areas) of this grass were collected using 1 m² quadrats located along the same transects used to measure fever-tree distribution ($N = 73$ plots). The number of available grass seed heads per quadrat was counted ($N = 13$), and the area covered by this grass was estimated to the nearest half-meter in these same areas. I used the average

**Table 3–7** Nutrients and Toxins of Foods (Percentage of Dry Weight Unless Indicated)

| Species and Source | Protein | Moisture | Fiber | Ash | Lipid | Tannin | Carbohydrate |
|---|---|---|---|---|---|---|---|
| *Acacia mellifera* flowers± (6) | 35 | 76 | 10 | | 8 | 2.3 | |
| *A. nilotica* flowers± (6) | 13 | 61 | 12 | | 4 | 3.0 | |
| *A. etbaica* flowers± (6) | 19 | 69 | 11 | | 9 | 1.9 | |
| *A. tortilis* flowers± (4, 6) | 19, 21 | 74, 79 | 17, 17 | 6 | 7, 10 | 4.5 | 46 |
| *A. xanthophloea* flowers (4, 7, 9) | 5, 1*, 1* | 75, 75, 7*, 4* | 18, 4*, 5* | 7, 1*, 2* | 5, 1*, 1* | 3.4 | 65, 19*, 16* |
| *A. drepanolobium* new growth (8) | 15 | | 30 | 6 | | | |
| *A. seyal* new growth (8) | 15 | | 23 | 10 | | | |
| *A. seyal* leaves (8) | 15 | | 11 | 7 | | | |
| *A. xanthophloea* leaves (8) | 16 | | 18 | 8 | | | |
| *A. xanthophloea* new leaves (4, 7, 9) | 21, 6* | 70, 72* | 14, 4* | 7, 2* | 3, 1* | 3.2 | 56, 16* |
| *A. seyal* seedpods (6, 8) | 29, 22 | 71 | 13, 27 | 6 | 4 | 2.3 | |
| *A. xanthophloea* seeds (7, 9) | 11* | 75* | 3* | 2* | 3* | 4.2 | 7* |
| *A. seyal* seeds (5, 6) | 49 | 74 | 5 | | 8 | 0.0, 0.8 | |
| *A. drepanolobium* gum (1, 2) | 7 | 10* | | 2 | | | 91 |
| *A. xanthophloea* gum (3, 7, 9) | 1–2, 1* | 11–13*, 25* | 1* | 2–6, 2* | 0.2* | 0.2–0.3 | 92–97, 71* |
| *Lycium europaeum* leaves (4, 5, 6, 7) | 36, 39† | 90†, 85 | 23, 14, 12† | 20† | 6† | 0.2, 0.0 | 23† |
| *L. europeum* new leaves (9) | 4*, 7* | 90*, 82* | 1*, 3* | 2*, 2* | 1*, 1* | | 2*, 5* |
| *Asparagus* new leaves (8) | 14 | | 40 | 8 | | | |
| *A. africanus* leaves (6) | 31 | 79 | 14 | | | 0.0, 0.0 | |
| *Pennisetum* leaves (6) | 27, 32 | 83, 74 | 27, 27 | | 8 | 0.0 | |
| *Ipomoea* flowers (6) | 16 | 88 | 12 | | 12 | 0.7 | |
| *Grewia tembensis* new growth (8) | 9 | | 33 | 8 | | | |
| *G. tembensis* green fruit (6) | 13 | 85 | 47 | | | 1.3 | |
| Mushroom species (6) | 9 | 68 | — | | | 0.0 | |
| Caterpillar species (6) | — | 86 | — | | | 0.0 | |

Percentages may not add to 100 as not all variables in columns are mutually exclusive.
*Wet weight; ±foods not included in Segera primates' diets but listed for comparison;
†percentages from Klein (1978) are for immature leaves.

Sources: (1), Anderson and Dea (1967); (2), Anderson (1978); (3), Anderson and co-workers (1984); (4), Klein (1978); (5), Barton and Whiten (1993); (6), Barton and co-workers (1993); (7), Wrangham and Waterman (1981); (8), Dougall, Drysdale, and Glover (1964); (9), Altmann, Post, and Klein (1987).

dimensions of the grass patches recorded along transects within the vervets' home range to estimate the area within the riverine habitat covered during the wet season. Grass areas averaged 9 m in width ($N = 13$). Using these data and the mean number of grass seed heads available to vervets per square meter, I estimated the proportion of the vervets' range consisting of this grass and the number of seed heads available within their home range. During wet-season months, this area was estimated to be approximately 18,600 m², or 16.5% of the riverine habitat used by vervets and 4.7% of their entire home range. On average, 135 *B. brizantha* seed heads were available per square meter in these areas, or over 2,500,000 seed heads within their entire home range.

Juvenile vervet feeds on *Brachiaria brizantha* grass seeds. *(Photo by the author)*

I analyzed the distributions of the different terrestrial herbaceous vegetation (THV) foods to assess whether they differed from random. When all herbaceous-level food plants, including grasses, were analyzed, their distribution in the home range of patas monkeys did not differ significantly from random (difference between variance and mean number of food plants: $\chi^2 = 48.068$, df $= 59, p > .80$). The distribution of THV (including grasses) within the whistling-thorn woodland habitat of the home range of vervets was, however, nonrandom ($\chi^2 = 86.249$, df $= 52, p < .01$). Because the variance exceeds the mean (mean number per m$^2$ = 18.9 ± 31.4), vervet foods can be classified as clumped or aggregated rather than randomly or uniformly distributed (Ludwig & Reynolds 1988). The distribution of nongrass THV within the patas monkeys' range also was random, whereas the distribution of nongrass THV within the vervets' range differed significantly from a random distribution ($\chi^2 = 93.334$, df $= 53, p < .001$). Given the high degree of variance from the mean (mean number per m$^2$ = 11.4 ± 20.1), nongrass THV foods within whistling-thorn habitat in the vervets' range can be described as spatially clumped (Ludwig & Reynolds 1988). The number of feeding-plant species available within meter-square quadrats in the whistling-thorn woodland used by the two monkey species did not differ ($f = 0.209$, df $= 1, p = .649$). When grasses were excluded, because these were rare in the diet of both species (except for *B. brizantha*) in whistling-thorn habitat, an ANOVA revealed that nongrass food plants' densities also were similar in vervet and patas monkeys' ranges ($f = 0.498$, df $= 1, p = .482$).

Based on the abundance, distribution, nutritional quality, and processing costs of the different whistling-thorn food types, I constructed tables that rank the contestability of these different foods on an ordinal scale (Tables 3–8 and 3–9). Nutritional quality ratings are based on a combination of values for protein, fiber, sugar, mineral, carbohydrate, and toxin content. Processing costs ratings are based on whether an item is protected by obligate ant species as well as the feeding rates exhibited by vervet and patas monkeys on the different food types. Foods that were processed

**Table 3–8** Ratings of Different Vervet Food Types in Ascending Order from Least to Most "Contestable"

| Food Type | Nutritional Quality | Processing Costs | Abundance | Distribution | Relative Rank | Overall Rank (Mean) |
|---|---|---|---|---|---|---|
| Gum | 4 | 3 | 1 | 1 | 1 | 2 |
| Swollen thorns | 1 | 5 | 5 | 5 | 5 | 4.2 |
| New growth | 2 | 3 | 3 | 5 | 4 | 3.4 |
| Seeds | 2 | 2 | 4 | 1 | 3 | 2.4 |
| Flowers | 5 | 1 | 2 | 1 | 1 | 2 |

**Table 3–9** Ratings of Different Patas Monkey Food Types in Ascending Order from Least to Most "Contestable"

| Food Type | Nutritional Quality | Processing Costs | Abundance | Distribution | Relative Rank | Overall Rank (Mean) |
|---|---|---|---|---|---|---|
| Gum | 4 | 1 | 2 | 2 | 1 | 2 |
| Swollen thorns | 1 | 5 | 5 | 5 | 5 | 4.2 |
| New growth | 2 | 2 | 3 | 5 | 3 | 3 |
| Seeds | 2 | 5 | 4 | 2 | 4 | 3.4 |
| Flowers | 5 | 2* | 1 | 1 | 1 | 2 |

*No data are available on feeding rates on flowers by patas monkeys; processing cost rating is based on lack of ant protection and assumptions based on vervets' feeding rates on flowers.

more quickly are considered to incur lower processing costs than foods that were processed more slowly (see Chapter 4). Abundance is based on the number of items available per hectare, whereas distribution is based on the location of items in space. For example, gum is not found on trees of all heights, so it is considered more scarce and thus more likely to be contested than food items such as swollen thorns, which are found on trees of all heights. I produced an overall relative rank by averaging the ratings of a food item according to the different variables. Later, I compare these predictions with data on contests of vervet and patas monkeys, and consider the importance of the different variables contributing to the contestability of a food.

### Food Availability on Segera

The main food species of patas monkeys and vervets (Isbell 1998; Pruetz & Isbell 2000) varied in availability to the monkeys. Compared with whistling-thorn trees, fever trees are larger, individuals are more widely spaced, and density is less than that of whistling thorn. Using Whitten's (1988) definitions of patch size for vervets, individual fever trees and whistling-thorn trees would be considered small feeding sites (<10 m crown diameter). Each would be expected to incite within-group competition according to van Schaik's (1989) model because not all monkeys could feed simultaneously. Usurpability of fever-tree foods compared with whistling-thorn foods is more likely as well because the defense by obligate ant species of whistling-thorn results in ant aggression that was hypothesized to deter subsequent foragers. In other words, ant aggression in whistling-thorn patches would result in a short window of time during which a forager could feed in a patch relatively undisturbed by ants. Additionally, fever trees are relatively more widely distributed and provide larger feeding patches (assuming that crown size scales with food abundance in fever trees as it does with whistling-thorn *Acacia*). Thus, fever trees can be viewed as a food source that is patchy compared with whistling-thorn trees. In support of this conclusion are Isbell and co-workers' (1998b) findings that adult female vervets during the same study traveled farther between feeding sites in whistling-thorn compared to riverine habitat. These authors concluded that such a pattern was indicative of less-abundant feeding sites or sites with short food-site depletion times within whistling-thorn woodland (Isbell et al. 1998b). *Scutia myrtina* shrubs provide another relatively large and widely dispersed food patch for vervets in the riverine habitat, although one that is available seasonally.

The availability of foods at a different vegetative level than trees or shrubs was also considered. The density of food items in the herbaceous level did not differ within the home ranges of the two primate species whether they were considered together or according to the separate species. Certain methodological problems were apparent upon examination of the results stemming from these data, however. For example, the food plant species *Solanum incanum, Lycium europeum,* and *Tribulis terrestris* were not recorded in quadrat samples in the case of vervets, although they did occur in the vervets' home range. Patas monkey quadrats lacked *Asparagus* species and *Bulbine abyssinica* food plant species, although these species also occurred within the patas monkeys' home range. Byrne and co-workers (1990) found that strip plots used to estimate plant food densities for baboons similarly did not sample all foods included in the diet. Although Byrne and co-workers (1990) concluded that strip plots were not adequate measures, in my study this method sampled the majority of food plant species included in the diets of monkeys but with some important exceptions.

The patterns of distribution of herbaceous vegetation between the home ranges of vervets and patas monkeys did differ. Although herbaceous-level foods were randomly distributed within the patas monkeys' range, these same foods were clumped within the vervets' range in the whistling-thorn woodland habitat. This aspect of food availability should therefore be considered when comparing contest competition between the two primate species in whistling-thorn woodland because clumped foods are predicted to elicit high levels of contest competition (Isbell 1991; van Schaik 1989; Sterck, Watts, & van Schaik 1997; Wrangham 1980). Additionally, foods available to vervets in the riverine habitat are more abundant and clumped in space (i.e., larger feeding sites more widely distributed in space) than those in whistling-thorn woodland. *Brachiaria* grass accounted for more than 16% of the riverine habitat used by vervets.

The lack of standardized methods used to measure food availability in most studies makes it difficult to compare results across sites. Nonetheless, the abundance of whistling-thorn trees within the woodland habitat on Segera is striking when compared with the density of major feeding tree species from other East African sites. Whitten (1988) estimated that

*A. tortilis* trees occur at a mean number of 1.8 per hectare within the home range of vervets in northern Kenya, with a total of 108 reproductively mature *A. tortilis* trees found within vervets' 60 hectare home range. Considering that approximately 29 hectares of the vervets' range on Segera consists of whistling-thorn woodland, over 71,000 whistling-thorn trees are estimated to be available to the study group. Although *A. tortilis* is larger than whistling-thorn *Acacia* and comparable in size to fever trees, even fever trees are abundantly available (57 per ha) for vervets on Segera in comparison to the availability of *A. tortilis* at the Samburu site. Estimated whistling-thorn density for patas on Segera exceeds 4 million trees.

Comparisons between sites are limited for a number of reasons. Crown volume is extremely small for whistling-thorn feeding trees, especially in comparison with typical feeding trees of forest species and even other *Acacia* species. Additionally, the amount of food within a crown was quantified through counts for whistling thorn but is rarely, if ever (but see Chapman et al. 1992), quantified at a more specific level than crown size in other studies. The size of food items, in addition to their amount, within feeding species' crowns should also be taken into account, as well as the nutritional quality of foods. Researchers have made attempts to make valid between-site comparisons for savanna baboons (Barton, Byrne, & Whiten 1996) but rarely for other primate species. The results of this study do not necessarily indicate that monkeys on Segera have absolutely more food available than other primates but that the number of feeding trees or patches available to them is large. The potential impact of such a large number of feeding sites of the same food species on vervet and patas monkey social behavior is examined in subsequent chapters.

# 4

# Feeding Behavior of Patas and Vervet Monkeys

A significant component of my study entailed examining the feeding behavior of vervets and patas monkeys. Because species differences in processing foods or in diet composition could result in differences in monkeys' perception of the availability of the same foods, they could ultimately influence female contest competition over food. Variables such as body size and food-processing apparatus (teeth and guts) constrain a primate's feeding behavior. Subsequently, differences in feeding behavior between species may be related to physiological and morphological differences (Richard 1985). For example, the African and Asian colobines possess a sacculated stomach that aids in fermentation of food as well as providing considerable volume to allow a large amount of food to be processed. This trait permits the colobines a largely folivorous and, presumably, low-quality diet. Accurately assessing primate food availability in the absence of information on how primates interact with food sources does little to advance theories of primate behavior and ecology. Therefore, in order to understand feeding competition, detailed data on primate feeding and foraging behavior are necessary (Janson 1988).

## STUDYING THE FEEDING BEHAVIOR OF SYMPATRIC SPECIES

The premise of comparing two closely related primate species in the same environment is that differences in ecological variables are minimized. In other words, each species potentially has access to the same general set of resources. Subsequently, behaviors traditionally linked to ecology, such

**67**

as female contest competition over food, can be interpreted as resulting from the influence of patterns of food availability in the environment. However, the allowance must also be made that, although habitats seem similar, the perception and use of the environment by individuals and species ("umwelt" or "self-world," Clemmons & Buchholz 1997) may differ. Understanding a primate's perception of and interaction with its environment is crucial for interpreting the socioecology of a species. To put it simply, assessing the range of feeding options available to an individual is essential when addressing questions of the effects of food availability on primate behavior.

Differences in feeding and foraging behavior among primate species may lead to differences in the way primates perceive foods (e.g., whether they are usurpable). Moreover, the same individual may perceive or rank the same patch differently depending on factors such as motivation, satiation, time since last feeding, change in reproductive condition, and number of conspecifics in the immediate area. Variation in feeding behavior can occur within as well as between populations and may reflect cultural traditions as well as patterns of food availability (McGrew 1992; Wrangham 1977). For example, blue monkey (*Cercopithecus mitis*) groups in Kibale Forest, Uganda, differed in the number of food plant species they shared in common, even though the two study groups ranged within 500 m of one another (Rudran 1978). Only 45% of all food plant species ($N = 74$) used by blue monkeys were eaten by both groups, and only 18% of these were included in the top twenty food plant species of each group (Rudran 1978). Additionally, shared food plant species were used at different frequencies by the groups. Some food plant species were available in higher densities in one study group's home range and were correspondingly used more. Other species, however, were preferred by one or other of the groups regardless of their availability (Rudran 1978). If differences in feeding behavior lead to differences in primate species' competitive behavior, divergence in feeding behavior may also lead to disparate dominance styles among primates (Sterck & Steenbeek 1997).

Differences in diet may influence contest competition and, consequently, dominance styles within a primate species (Isbell 1991; Wrangham 1980). For example, Koenig and co-workers (1998) compared two forest populations of Hanuman langurs (*Presbytis entellus*) and found that Kanha langur females relied heavily on the superabundant *Shorea robusta* tree for a variety of food types, and adult female dominance relations were "egalitarian" at this site. In contrast, at the Ramnagar site, where this same tree also was available (although not as abundant), female langurs used this feeding species much less, and the adult female dominance hierarchy was linear and stable (Koenig et al. 1998). Thus, differences in feeding behavior in conjunction with differences in food availability resulted in disparate female social relationships at the two sites. It should be noted, however, that food availability measures between the two sites were not consistent, and, therefore, it remains unclear as to exactly what specific aspects of food availability may have caused this effect.

Both patas monkeys and vervets on Segera exhibit a high degree of dietary overlap and use whistling-thorn woodland extensively, which is an atypical habitat for vervets in East Africa. Studies of vervets in East Africa have described this species as mainly florivorous (Klein 1978; Whitten 1982; Wrangham & Waterman 1981), whereas patas monkeys in East Africa are insectivorous and gumnivorous (Chism & Rowell 1988; Isbell 1998). Given the close degree of relatedness between vervets and patas monkeys as well as their behavioral similarities (e.g., adaptations related to woodland and woodland-savanna habitats), these primates are expected to exhibit similarities in feeding behavior and intragroup social behavior when using the same habitat type.

Because neither vervets nor patas monkeys exhibit dietary specializations unique to their subfamily (Cercopithecinae) or tribe (Cercopithecini), I expect differences in diet exhibited by these species when using the same habitat (i.e., whistling-thorn woodland) to be caused by other factors such as specific ecological variables that differ for the two species. All Cercopithecids possess bilophodont molars, which, among the high-cusped colobines, serve to shear foods, and, among the low-cusped cercopithecines, serve to crush foods (Richard 1985). Cercopithecines are also characterized by having cheek pouches that function to store food and assist in predigestion.

In my study, I quantified both the abundance and distribution of foods used by patas and vervet monkeys at the Segera site and bring together this information with data on the feeding behavior of these species. I examined monkeys' interactions with their main food species to assess the interrelationships between aspects of primate behavior (feeding behavior, feeding competition, and dominance relationships) and ecology (food abundance and distribution). Although vervet and patas monkeys on Segera used many of the same food items, and exploited similar woodland habitats, because they also exhibit significant differences in social organization, body size, and life-history traits, it might be expected that each species exploits the same food resources differently. For example, greater body size may enable patas monkeys to exploit foods that require processing more efficiently, such as the woody, ant-protected swollen thorns of whistling-thorn *Acacia*.

## PATAS AND VERVET MONKEYS: EXPECTED DIFFERENCES AND SIMILARITIES

Patas and vervet monkeys exhibit a number of morphological differences that could result in disparities between the species in feeding behavior.

**Figure 4–1** Subadult patas monkey foraging bipedally on *A. drepanolobium*. *(Photo by the author)*

These include body size, canine size, and limb length. Vervet females weigh from 2.5 to 5.3 kg (Haltenorth & Diller 1977), whereas adult female patas monkeys average 6 kg (Sly et al. 1983). Additionally, patas monkeys have elongated limbs relative to vervets (Hurov 1987; Strasser 1992). Both species often stand bipedally on the ground while feeding on defended whistling-thorn food items rather than feeding from within trees (Figure 4–1). This may serve to reduce the disturbance of defensive ants. Larger individuals (and those with greater bipedal height and arm reach) would be expected to have access to a greater number of food items while also better avoiding ant aggression. Larger absolute canine size is expected to be an advantage in opening the woody swollen thorns of this *Acacia*. Individuals often gain access to the insects by first puncturing the woody thorn with a canine and then prying open a hole using the lower incisors and canines to form a flap-like opening. Efficient feeding (fast feeding rates) on swollen thorns would be advantageous to avoid attack by the obligate ants.

Behavioral differences in avoiding ant attack between patas and vervet monkeys could also contribute to differences in feeding behavior between these species. Group spread in patas monkeys is large. Chism and Rowell (1988) found that one group of patas monkeys ranging in size from sixteen to forty-one individuals spread over a mean distance of 96 m, whereas another group that ranged from forty-seven to seventy-four individuals spread over a mean distance of 179 m. Foraging patas monkeys would therefore be less likely to encounter a previously disturbed whistling-thorn patch than vervets because the latter species forages more cohesively. Isbell and Enstam (2002) found that vervet groups on Segera were more wide-spread in whistling-thorn woodland than in riverine habitat. I used the distance from foraging individual to nearest neighbor to provide an esti-mate of social proximity in vervet and patas monkeys while they fed on whistling-thorn foods.

An important factor influencing feeding behavior in some primates, specifically regarding access to foods, is the dominance status of individu-als (Isbell 1991; Wrangham 1980, 1988). Because food resources influence females' reproductive success, where food is limited and variance between feeding sites is high, females are expected to contest foods and subse-quently form dominance hierarchies within a group (van Schaik 1989). The dominance hierarchy among female vervets has been reported as linear and stable (Horrocks & Hunte 1983; Isbell & Pruetz 1998; Seyfarth 1980; Whitten 1982), whereas the dominance hierarchy of free-ranging, nonpro-visioned patas monkeys has been reported to vary over time (Isbell & Pruetz 1998). Individual differences in feeding behavior are thus predicted to be greater for female vervets compared to patas monkeys because rank differences are expected among the former. Dominant female vervets should exclude subordinate females from food patches if those patches are worth usurping.

The aggressive behavior characteristic of the ants obligate with whistling-thorn *Acacia* led me to predict that both patas and vervet mon-keys' feeding behavior would be influenced by the ants' defense of certain whistling-thorn food items. Stapley (1998) showed that ant defense, in combination with thorn defense, significantly reduced herbivore brows-ing on whistling-thorn trees in Tanzania. Ants' defense of particular food items was predicted to affect monkeys' foraging, feeding, and ranging behavior because individuals are unlikely to exploit all food items on a single tree, especially given the abundance of whistling-thorn patches in the home ranges of monkeys on Segera. I hypothesized that ant protection would affect the monkeys' feeding behavior by forcing them (1) to spend less time feeding per patch, thus exploiting a larger number of whistling-thorn patches rather than depleting a few patches, (2) to feed on defended food items proportionately less in relation to their apparent availability than nondefended food items, and (3) to feed in whistling-thorn food patches individually rather than in groups of more than one animal in order to reduce ant disturbance. Each of these behavioral mechanisms would allow primates to more efficiently avoid ant attack.

In summary, factors such as morphology and anatomy, physiological limitations such as those associated with digestion, processing food costs (e.g., associated with ant protection), dominance rank of individuals, and feeding preferences should be considered when examining the feeding ecology and related behaviors (i.e., competition and dominance) of pri-mates. I focused on the food species whistling-thorn *Acacia* because it was important in the diet of both primate species studied. The nature of this food plant species also allowed me to take absolute measurements of food availability not normally encountered in a field situation.

## WHISTLING-THORN *ACACIA:* WHY FOCUS ON ONE FOOD SPECIES?

*Acacia* is an important food plant in the diets of woodland-savanna and savanna-dwelling primates such as vervets, baboons, and patas monkeys. In Uganda, leaves of *Acacia* were one of the principal components of the diet of vervets and baboons, and vervets contested this resource (Gartlan & Brain 1968). *Acacia xanthophloea, A. drepanolobium,* and *A. tortilis* have been cited as important vervet food species at several sites in East Africa (Isbell et al. 1998b; Whitten 1983; Wrangham & Waterman 1981). Studies at the Mutara ADC site in Kenya found that whistling-thorn *Acacia* accounted for about 85% of patas monkeys' feeding time (Chism & Wood 1994). *Acacia* is an important food for both patas and vervet monkeys on Segera. Whistling thorn made up 83% of the diet of the group of patas monkeys studied here over 17 months, from 1992 to 1994 (Isbell 1998). Whistling thorn and other species of *Acacia* (*A. xanthophloea* and *A. seyal*) accounted for >40% of the vervet study group's diet from July 1993 to December 1994 (Isbell et al. 1998b).

Given the importance of foods of *Acacia* in the diet of monkeys on Segera, its availability is expected to affect feeding behavior, and subsequently, contests over food and dominance behavior among adult females. This has been documented in Japanese macaques, where contests between adult females varied with the patch size of their main food item, seeds of *Zelkova* (Saito 1996). Females interacted agonistically more often in isolated (distance to neighboring patch >30m), small (containing less than three feeding sites) patches than in large ones (Saito 1996). The nature of whistling-thorn *Acacia*—the small tree rarely grows to heights exceeding 4 m—allowed me to quantify food availability of feeding trees on a number of different scales and to a degree not usually possible in a field situation.

## FEEDING BEHAVIOR OF VERVETS AND PATAS MONKEYS

In measuring feeding behavior, I combined eating and foraging into one category that includes ingesting, manipulating, and searching for foods. The following terms were used, and 1–3 are defined here after Post (1982: 408):

(1) Food type—a specific plant part irrespective of species, e.g., swollen thorns.
(2) Food species—a plant species, irrespective of food type, e.g., whistling-thorn *Acacia*.

(3) Food item—a combination of food type and species, e.g., whistling-thorn gum.

(4) Food site—area where monkey forages or feeds while stationary.

## Continuous Sampling

I used continuous and bout sampling (Altmann 1974) of focal individuals to record feeding behavior and habitat use. Besides subject's identity, time of day, and date, data that I collected during 30-minute continuous sampling periods included: (1) time spent feeding/foraging or in other activities, (2) the number of whistling-thorn trees/patches fed on, (3) the number of feeding sites used and food items eaten, and (4) the time spent in different habitat types. The obligate *Acacia* ant species resident on individual feeding trees was also recorded when possible. An attempt was made to observe a different subject each hour from 0800 to 1500 hours four times per month from January to December 1994. Eight adult female vervets were observed on a predetermined random schedule, which was part of Lynne Isbell's research protocol, so that each subject was observed once per day at a different time of day during the same month. A similar schedule was abandoned for adult female patas monkeys because they were more difficult to identify and locate. However, I made an attempt to sample female patas monkeys once per day at different times of the day during the same month.

A total of 435 focal animal follows were analyzed regarding the feeding behavior of female vervets ($N = 194$) and patas monkeys ($N = 241$) on whistling-thorn *Acacia*. Patas monkeys fed approximately 39% of the time they were observed. Vervets fed proportionately more in whistling-thorn woodland, and feeding accounted for 76% of observation time there. In riverine habitat, vervets spent only 31% of their time feeding (Pruetz 1999; Pruetz & Isbell 2000). Both monkey species were similar regarding the number of feeding trees, feeding sites, and feeding sites per tree used during a sample period (Table 4–1). Vervets fed in fewer fever trees than whistling-thorn trees, at more fever tree than whistling-thorn feeding sites, and at more fever tree sites per tree per unit time, however.

## Bout Sampling

To explore specific aspects of foraging behavior, feeding bouts by monkeys on whistling-thorn *Acacia* were recorded during each of 18 months from September 1993 to February 1995. Observations were conducted opportunistically during each hour of the day between 0730 and 1600 hours when continuous behavioral data were not being collected. Scans of group members were conducted, and the first individual seen approaching a whistling-thorn feeding site was sampled. Feeding bouts were defined

**Table 4–1**  Use of *Acacia* Feeding Trees (Mean Per 30-Minute Sample)

|  | *Vervets* | *Patas Monkeys* |
|---|---|---|
| Number of whistling-thorn trees fed in per sample | 4.7 ± 8 (N = 214) | 5.0 ± 4.7 (N = 220) |
| Number of whistling-thorn food sites per sample | 6.0 ± 9.6 (N = 214) | 6.7 ± 6.4 (N = 220) |
| Number of whistling-thorn food sites per tree | 1.3 ± 0.5 (N = 116) | 1.4 ± 0.7 (N = 194) |
| Number of fever trees fed in per sample | 1.1 ± 1.2 (N = 214) | 0.01 ± 0.11 (N = 220) |
| Number of fever tree food sites per sample | 5.9 ± 9.2 (N = 214) | 0.02 ± 0.17 (N = 220) |
| Number of fever tree food sites per tree | 5.8 ± 6.7 (N = 130) | 1.3 ± 0.6 (N = 3) |

as eating or foraging in a single whistling-thorn patch. A patch was defined as an area where an individual could continuously eat while moving (after Chapman 1988 and White & Wrangham 1988). For example, the crown of one fever tree *Acacia*, one whistling-thorn *Acacia*, or overlapping crowns of whistling-thorn trees could constitute a patch. In most cases, monkeys fed at a single tree within a patch, and one to four whistling-thorn trees with contiguous canopies constituted a single patch (Pruetz 1999). A feeding bout began when a focal subject began manipulating or actively searching within a patch. A bout ended when the focal subject directed its attention away from the patch by leaving the vicinity of the patch, turning or visually orienting away (e.g., Post 1982) from the patch and discontinuing foraging. A total of 568 feeding bouts were recorded (271 patas monkey and 297 vervet samples). Data analyzed include (1) subject's identity, (2) date, (3) time of day, (4) duration (in seconds) of feeding bout, (5) number of items consumed, (6) food item type (swollen thorns, new growth, gum, seeds, flowers, and non-ant insects), (7) focal subject's nearest neighbor, (8) distance to nearest neighbor, (9) whether nearest neighbor was feeding and/or within same patch as focal subject, and (10) obligate ant species resident within patch. Intraobserver reliability in estimating distance in meters was tested each month as part of Lynne Isbell's research protocol by estimating a number of horizontal distances between points and calculating percentage of accuracy with actual distances. I scored greater than 95% reliability in eleven (N = 40–80 per month) samples from February to December 1994. Ecological data gathered on feeding trees included number of trees in a patch, crown height of patch, and two maximum patch diameter measures (one perpendicular).

Values for each individual adult female vervet were summarized, and the mean value for each female was used in analyses of feeding bout data. Data on one vervet female (MND) were not used because only one sample bout was available for her, and she disappeared early in the study. Data on individual patas monkey females were pooled owing to the small sample size available for most individuals. Pooling data may obscure individual variance as well as overweight dependent samples, however. For example, individual female patas monkeys varied in the number of items fed on per bout on whistling thorn (ANOVA: df = 13, $F$ = 1.58, $p$ = 0.005). This difference was owing to one outlier (GEO), and when she was excluded from analyses, individuals' values were similar (ANOVA: df = 12, $F$ = 0.44, $p$ = 0.945). Because variation in individual females' feeding behavior was otherwise insignificant for patas monkeys (i.e., regarding feeding rates, nearest neighbor distances, bout duration, feeding height, and patch width— see Pruetz 1999), data were subsequently pooled for analyses. Calculations of mean feeding rates (defined here as the average number of seconds it took to process and eat one food item) of the different types of whistling-thorn foods were conducted on pooled values for adult female vervets and patas monkeys owing to the small sample sizes for individuals regarding the different food types (Table 4–2). Pooled and unpooled means of the proportions of different whistling-thorn foods used during feeding bouts by adult female vervets were similar (Pruetz 1999; Table 4–3).

Vervet and patas monkey females exhibited many similarities when feeding on the same food species (see Table 4–2). They were similar in the duration of time they spent feeding per bout on whistling-thorn food items, both defended and nondefended food types, the number of items eaten per bout, distance from focal subject to nearest neighbor, and maximum patch width. The feeding rate on all whistling-thorn foods and on defended food items differed significantly between the species, however. Patas monkey females fed, on average, slower than adult female vervets (see Table 4–2). Time of day had no effect on the feeding bout duration of either species (Pruetz 1999). The behavior of focal subjects' nearest neighbors during feeding was also examined. Rarely was the nearest neighbor

**Table 4–2** Female Vervet and Patas Monkey Feeding Behavior on Whistling-Thorn *Acacia**

| *Mean Values* | *Vervets* | *Patas Monkeys* | *t-Value* | *p-Value* |
|---|---|---|---|---|
| Food items eaten per bout | 2.6 | 2.1 | 1.49 | 0.138 |
| Height of food items | 0.65 m | 0.70 m | −1.807 | 0.072 |
| Patch width | 2.3 m | 2.6 m | −0.557 | 0.578 |
| Nearest neighbor distance | 7.4 m | 9.7 m | −1.475 | 0.141 |
| Bout duration (ant-defended foods) | 18.5 s | 20.5 s | −0.583 | 0.561 |
| Bout duration (all foods) | 22.5 s | 23.7 s | −0.312 | 0.756 |
| Feeding rate (all foods) | 8.5 s | 11.1 s | −2.829 | 0.005 |
| Feeding rate (defended foods) | 9.2 s | 12.2 s | −3.255 | 0.001 |

*Independent *t*-tests.

**Table 4–3** Feeding Behavior on Whistling-Thorn by Female Vervets According to Rank

| Behavior | High Rank | Low Rank | p-Value | F-Value | Sample Size |
|---|---|---|---|---|---|
| Bout duration | 18.4 ± 31.8 s | 26.9 ± 27.7 | 0.088 | 2.958 | 146 |
| Feeding rate | 7.9 ± 4.8 s | 9.7 ± 5.2 | 0.058 | 3.657 | 146 |
| Defended bout duration | 13.8 ± 9.7 s | 25.8 ± 28.0 | 0.003* | 8.964 | 109 |
| Defended food rate | 8.6 ± 4.8 s | 9.9 ± 4.5 s | 0.150 | 2.100 | 109 |
| Number of items per bout | 3.0 ± 6.5 | 3.0 ± 2.9 | 0.908 | 0.013 | 148 |
| Food height | 0.6 ± 0.3 m | 0.6 ± 0.4 m | 0.720 | 0.129 | 112 |
| Tree height | 1.0 ± 0.6 m | 1.0 ± 0.4 m | 0.951 | 0.004 | 137 |
| Number of trees in patch | 4.5 ± 5.6 | 3.0 ± 3.7 | 0.097 | 2.795 | 116 |
| Patch width | 2.5 ± 2.2 m | 2.3 ± 1.4 m | 0.615 | 0.254 | 107 |
| NN distance | 7.2 ± 5.4 m | 8.8 ± 7.5 m | 0.188 | 1.753 | 134 |

*Denotes significant difference.

feeding in the same patch as the focal animal patas monkey or vervet (4% and 1.5% of bouts, respectively), even though the nearest neighbor was often feeding (42% and 65% of bouts, respectively).

Differences between the monkey species were evident in the proportion of different foods eaten (as measured by number of items eaten). Patas monkeys most often ate insects within swollen thorns whereas vervets fed on new growth most frequently (Figure 4–2). Sixty-five percent of all feeding bouts by patas monkeys and 74% of all feeding bouts by vervets on whistling-thorn included foods that were defended by ants, such as new growth and swollen thorns.

Analyses of variance tests revealed that there were significant differences in the rates at which the primate species processed different types of whistling-thorn food items (patas monkeys: df = 4, $F$ = 2.604, $p$ = 0.036; vervets: df = 5, $F$ = 3.854, $p$ = 0.002). Vervets processed flowers most quickly (4 s per item) and gums most slowly (8 s per item). For patas monkeys, non-ant insects on whistling-thorn were most quickly processed (5 s per item), and seeds were processed most slowly (17 s per item).

As predicted, the type of ant species resident on whistling-thorn trees significantly affected monkeys' feeding behavior. Vervets' feeding behavior on defended foods varied according to the obligate ant species resident regarding the duration of feeding bouts (ANOVA: df = 3, $F$ = 4.59, $p$ = 0.004) and the number of food items eaten per bout (ANOVA: df = 3, $F$ = 2.76, $p$ = 0.042). Vervets fed longer per bout and ate more defended food items per bout on trees that had *Lepisota* species resident compared to those that had *Tetraponera* and *Crematogaster* species resident. Patas monkeys' feeding behavior on defended food items differed significantly according to ant species regarding bout duration (ANOVA: df = 3, $F$ = 3.02, $p$ = 0.032), feeding rate (ANOVA: df = 3, $F$ = 3.28, $p$ = 0.023), and the number of

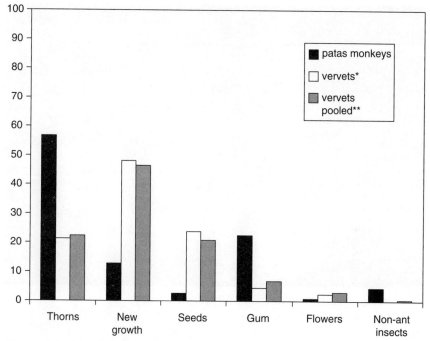

**Figure 4–2** Adult female feeding behavior on whistling-thorn food types.
*Data for individual female vervets, averaged.
**Data averaged for all female vervets.

defended food items eaten per bout (ANOVA: df = 3, $F$ = 4.28, $p$ = 0.006). Patas monkeys fed longer and faster when *Lepisota* species were resident and fed on more food items where *Tetraponera* species were resident. As I noted in the previous chapter, both *Lepisota* and *Tetraponera* species were less aggressive than *Crematogaster* species.

## RANK-RELATED DIFFERENCES IN FEEDING BEHAVIOR

In the next chapters, I discuss the dominance relationships of vervets and patas monkeys. Here, I use those data to classify adult female vervets according to rank in order to examine whether females of differing rank varied regarding their feeding behavior. When female vervets were classified as being one of four high- or four low-ranking individuals in a dominance hierarchy, differences according to rank were apparent (see Table 4–3). Low-ranking females fed significantly longer per bout on defended whistling-thorn food items. High-ranking females tended to have faster feeding rates than low-ranking females, although this difference only approached the set significance level.

Low-ranking vervet female "Mooshu." *(Photo by the author)*

High-ranking vervet female Frijole feeds on whistling-thorn *Acacia*. *(Photo by the author)*

Subordinate female vervets ate at more feeding sites per tree/patch ($F = 49.134$, df $= 1$, $p = 0.000$) and at more sites per individual tree than did high-ranking females (ANOVA: $F = 22.18$, df $= 1$, $p < 0.001$, $N = 92$). The number of feeding sites fed in by high- versus low-ranking vervet females in fever trees also differed significantly (ANOVA: $F = 4.635$, df $= 1$, $p = 0.034$, $N = 108$), with low-ranking females feeding in more sites per tree (see also Pruetz 1999).

## Patch Depletion

Analyses of vervet and patas monkey feeding behavior suggest that ant defense influences primate feeding behavior. Both patas and vervet monkeys fed on only a few food items per tree (Table 4–4), although many items were seemingly available per tree, and each fed longer at patches with less aggressive species resident. It seems plausible to explain short foraging times in whistling-thorn food patches as a result of ant protection of foods. However, to explore the possibility that short patch feeding times could be influenced by some characteristic other than ant defense, I analyzed the behavior of monkeys to see if they were depleting the food patches they used. Results from a study of howling monkeys (*Alouatta palliata*) and capuchins (*Cebus capucinus*) in Costa Rica suggested that these primates might deplete their food patches (Chapman 1988). Indicators of food patch depletion reported included: (1) a faster food intake rate at the beginning of feeding bouts in a single patch than at the end of bouts, (2) a decrease in feeding rates (seconds per item spent foraging/feeding) for all plant types during bouts, and (3) a positive correlation between time spent feeding in a patch and patch size (Chapman 1988). The possibility that monkeys depleted

**Table 4–4** Results of ANOVA Tests: Feeding Rate per Food Item According to Bout Length

| Food Type | Patas Monkeys | | | Vervets | | |
|---|---|---|---|---|---|---|
| | F-Value | p-Value | N | F-Value | p-Value | N |
| Swollen thorns | 26.091 | 0.000* | 130 | 16.597 | 0.000* | 61 |
| New growth | 42.670 | 0.000* | 51 | 37.364 | 0.000* | 156 |
| Gums | 7.537 | 0.008* | 52 | 9.573 | 0.007* | 17[†] |
| Seeds | 0.342 | 0.575 | 10[†] | 10.589 | 0.002* | 51 |
| Flowers | Not applicable | Not applicable | Not applicable | 0.017 | 0.902 | 6[†] |
| Non-ant insects | 4.068 | 0.067 | 52 | Not applicable | Not applicable | Not applicable |

*Denotes significant difference; all significant results showed that slower feeding rates occurred during long bouts.
†Note small sample size.

the whistling-thorn patches they fed in was explored by: (1) examining the feeding rate of monkeys during long and short bouts, (2) examining feeding rates on different food types during long and short bouts, and (3) examining feeding bout duration in relation to patch size.

Because monkeys' feeding bouts on whistling thorn were short, rather than examining feeding rates at the beginning and end of feeding bouts (Chapman 1988), I scored bouts as either long or short, based on median values of bout length for each species (15.8 s—patas monkeys, 13.4 s—vervets). Feeding bout data were analyzed separately according to the different food types (swollen thorns, new growth, seeds, gum, flowers, free-living animals, or non-ant insects). Feeding bout data for flower feeding by patas monkeys ($N = 2$) and non-ant insect feeding by vervets ($N = 5$) were excluded from analyses owing to small sample sizes. Analyses of variance tests were used to examine the interaction between bout length, feeding rate, and food type (see Table 4–4). The feeding rate of monkeys during long and short bouts, feeding rates for different food types during long and short bouts, and feeding bout duration in relation to patch size were analyzed to assess whether monkeys were depleting food patches. Results of the analyses showed that patas monkeys fed more slowly during longer bouts on swollen thorns, gum, and new growth. Vervets fed slower during longer bouts on swollen thorns, new growth, gum, and seeds but not on flowers.

Regression analyses on bout duration and patch size (as measured by tree height) were conducted according to food type (Table 4–5). In these analyses, data were pooled for all vervet age and sex classes, including adult males and immature vervets (see Pruetz 1999), because sample sizes regarding individual food types would otherwise be too small. The possibility that such pooling of data would obscure individual differences was tested using data on adult female vervets' feeding bouts. Individual female vervets did not differ significantly regarding their feeding rates on whistling-thorn foods (ANOVA: df $= 7$, $F = 0.604$, $p = 0.751$) or the average duration of whistling-thorn feeding bouts (ANOVA: df $= 7$, $F = 1.022$, $p = 0.418$). Adult females comprised approximately half of all feeding bouts analyzed ($N = 149$ of 296). Patas monkeys fed longer in larger patches when all food items were considered, when only non-ant defended foods were included in one category, and when feeding on new growth (see Table 4–5). Vervets fed longer in larger patches when all food items were considered, when only defended food items were considered, and when feeding on swollen thorns, new growth, and seeds (see Table 4–5).

## Monkeys, *Acacia*, and Ants

Primates on Segera visit many feeding trees in a single day. Considering the number of trees used per foraging minute and the amount of time spent foraging by adult females, patas monkeys fed in an estimated forty-six

**Table 4–5** Regression Analyses on Bout Duration as a Function of Patch Size

| Food Type | Patas Monkeys | | | | Vervets | | | |
|---|---|---|---|---|---|---|---|---|
| | R | F-Value | p-Value | N | R | F-Value | p-Value | N |
| Swollen thorns | 0.078 | 0.743 | 0.391 | 122 | 0.401 | 10.919 | 0.002* | 59 |
| New growth | 0.289 | 4.191 | 0.046* | 48 | 0.427 | 33.484 | 0.000* | 152 |
| Gum | 0.260 | 3.540 | 0.066 | 51 | 0.287 | 1.258 | 0.281 | 16† |
| Seeds | 0.511 | 2.822 | 0.131 | 10† | 0.453 | 11.639 | 0.001* | 47 |
| Flowers | Not applicable | Not applicable | Not applicable | Not applicable | 0.685 | 1.770 | 0.315 | 4† |
| Non-ant insects | 0.376 | 1.974 | 0.185 | 14† | Not applicable | Not applicable | Not applicable | Not applicable |
| Ant-defended foods | 0.113 | 2.179 | 0.142 | 169 | 0.416 | 43.813 | 0.000* | 211 |
| Nondefended foods | 0.303 | 7.363 | 0.008* | 75 | 0.227 | 3.529 | 0.065 | 67 |
| All foods | 0.322 | 28.185 | 0.000* | 246 | 0.365 | 43.295 | 0.000* | 283 |

*Denotes significant difference; all significant findings indicate longer feeding bouts in larger patches.
†Note small sample sizes.

Patas monkeys in whistling-thorn woodland. *(Photo by the author)*

whistling-thorn trees per day and at sixty-six whistling-thorn feeding sites per day. Vervets fed in an estimated sixty whistling-thorn trees per day and at seventy-eight whistling-thorn feeding sites per day. These results illustrate the ubiquity of whistling-thorn in this habitat and its importance as a food source to Segera primates.

Vervets and patas monkeys did not exploit individual whistling-thorn trees extensively. Both monkey species fed for short periods of time on a few food items in different individual whistling-thorn patches (Isbell 1998 & this study). Although trees contain many potential food items, both species fed on less than three items per tree, on average, indicating that ant attack restricts the amount of time feeding in and the number of items taken from each tree. Moving from one patch to another enables primates to avoid ant aggression. Other studies have also demonstrated the effective defense of obligate *Acacia* ants against predators of their host tree. In an experiment testing the effectiveness of thorns and ants as defenses against herbivores, Stapley (1998) found that thorns only slowed down feeding time, with herbivores compensating by feeding longer. Ant defense, however, in combination with thorn defense, caused browsing goats to stop feeding more quickly (Stapley 1998). In each of twenty experiments, browsing goats refused to return to whistling-thorn trees where they had been attacked by ants and instead moved on to forage at a different tree. Madden and Young (1992) found that young giraffes' feeding bout length correlated negatively with the number of swarming ants on a tree. They seem to employ the same strategy as the monkeys in this study, taking a few mouthfuls from a single tree and then moving on when the ants are disturbed (Dagg & Foster 1976).

Although ant defense does influence monkey feeding and foraging behavior on whistling-thorn *Acacia,* the possibility that additional variables could also affect these primates' interactions with this food species was examined. The possibility that monkeys were depleting whistling-thorn food patches was explored by examining feeding rates and bout durations. If feeding rate is an indicator of food abundance, analysis of variance tests indicated that both species depleted individual whistling-thorn patches during feeding bouts on certain items. Patas monkeys fed more slowly during longer bouts on swollen thorns, gum, and new growth. Vervets fed more slowly in longer bouts when feeding on swollen thorns, gum, new growth, and seeds.

If patches are not being depleted, feeding bout duration is expected to be independent of patch size (Chapman 1988). In other words, if primates are feeding on most available food sources in a patch, they are expected to feed longer in larger patches, on the assumption that larger patches have more food available. Tree dimensions, such as height and crown width, correlated with food abundance, with larger trees having more items available (see previous chapter and Pruetz 1999). Regression analyses showed that both species fed longer in larger food patches on some food items. Patas monkeys fed longer in larger whistling-thorn patches when non-ant defended foods were considered, but not when feeding on defended foods. This implies that feeding on defended food items is controlled by other factors (i.e., ant defense). When individual foods were considered, patas monkeys fed longer at larger patches when feeding on gum. Vervets fed longer in larger patches when feeding on swollen thorns, new growth, and seeds.

Based on the data analyzed here, it seems that monkeys are depleting individual whistling-thorn patches for particular food items. Patas monkeys fed longer in larger patches on gum and fed more slowly on gum during longer bouts. Vervets fed longer in larger patches on swollen thorns, new growth, and seeds and also fed more slowly on these foods during longer feeding bouts. Depletion of foods within a patch by monkeys is supported for these particular whistling-thorn food items but not for all foods. The primates may also perceive the aforementioned foods as depleted according to some variable or set of variables not discerned by a human observer. For example, gum varies in its consistency, and not all gum patches may be edible for patas monkeys. Tall trees (>2 m in height) in the patas monkeys' home range contain, on average, about nine gum patches per individual. Only half of these may be considered edible by patas monkeys (i.e., flowing or soft and pliable), so that feeding bout data on gum corresponds with the idea that patas monkeys deplete this food source. However, the implication that vervets deplete swollen thorn and new growth patches does not correspond with the availability of these foods. Although satiation could also explain slower feeding rates during longer bouts, the characteristically short feeding bout length and the use of multiple trees during feeding indicates that individuals are not satiated at most trees.

## Species Differences

Despite morphological and behavioral differences, vervets and patas monkeys exhibit many similarities when feeding on the same food resource. No significant differences between the two were revealed in the duration of feeding bouts on whistling thorn, the distance from focal subject to nearest neighbor during feeding bouts, number of food items eaten during bouts, and the heights of food items eaten. Using nearest neighbor distance as an indicator of within-group spread indicates similar spacing of individual patas and vervet monkeys while foraging in whistling-thorn woodland habitat. Both primate species were affected similarly by ant aggression when feeding, as indicated by bout duration and number of items eaten per bout. The height of feeding trees and feeding rates on defended food items such as swollen thorns and new growth, however, differed significantly between the species. The former difference can be explained by variation in the woodland profiles in the species' home ranges (see Figures 3–3 and 3–5). The latter finding can be attributed to differences in the way these primates used the whistling-thorn food source.

Vervets and patas monkeys differed in some respects regarding the composition of their diet when whistling-thorn foods were considered. Based on feeding bout data, patas monkeys ate more gum and animal prey from swollen thorns than vervets, who fed on new growth and seeds more than patas monkeys. Data from continuous focal animal samples of these same study groups during this time (January–December 1994) showed that patas monkeys fed on gums at 14% and on animals at 21% of feeding sites, whereas vervets fed on gum at 17% and on animals at 13% of feeding sites in whistling-thorn woodland (Isbell et al. 1998b). Vervets fed on seeds at 7% of feeding sites whereas patas monkeys fed on seeds at 1% of sites. These findings corroborate the data from vervet and patas monkey feeding bouts on whistling-thorn regarding animal prey and seeds but are in contrast with the bout data on gum feeding. Other data from this site indicate that patas monkeys fed on whistling-thorn gums 37% of the time (Isbell 1998), which does correspond with the data presented here from feeding bouts. Vervets' more efficient feeding (i.e., processing) on defended foods can be explained by the types of foods they commonly consume. Breaking off new growth is less time consuming than processing swollen thorns, underscoring the importance of considering processing costs in the analyses of feeding behavior and competition.

Processing costs differed for the primates according to the type of food eaten. Patas monkeys processed seeds most slowly whereas non-ant insects were processed most quickly. For vervets, gums were processed slowly whereas flowers were processed most quickly. Among yellow baboons in Amboseli, Kenya, processing time was the most important factor determining the success of feeding interruptions (Shopland 1987). Attempts to interrupt feeding conspecifics and resisting interruptions depended on feeding rates associated with particular food items (Shopland 1987). When assessing the costs of agonistic interactions among conspecifics, costs related to ant attack when feeding on defended whistling-thorn foods must be considered, as well as other aspects of these foods' availability. For example, although gums are associated with relatively slow processing times, they are patchily distributed compared to other whistling-thorn foods and thus hypothesized to be more monopolizable than other foods (see Figure 3–5). Additionally, gum is not defended, making it less costly to usurp.

## Within-Species Differences According to Rank

Rank-related differences among female vervets were expected because this species is reported to exhibit linear and stable dominance hierarchies (Cheney & Seyfarth 1990; Isbell & Pruetz 1998; Struhsaker 1968; Whitten 1983), and rank differences are assumed to result in individual differences in access to preferred food resources. For example, among male yellow baboons in Tanzania, high- and middle-ranking individuals ate fewer bites of foods with higher fiber content (Johnson 1990). Similarly, higher-ranking adult female olive baboons in Laikipia, Kenya, ingested more food (as measured by dry weight), more protein, and more fiber and also processed foods faster (Barton 1990). Barton (1990) concluded that low-ranking adult female baboons used inferior feeding sites compared to high-ranking females. In my study, low-ranking female vervets tended to feed longer per bout than high-ranking females on defended whistling-thorn foods and slower than high-ranking females when all whistling-thorn food types were considered. Low-ranking female vervets fed at significantly more sites per whistling-thorn patch, at more trees per unit time, at more sites per tree, and at more feeding sites per fever tree than high-ranking females. Like the baboons in Barton's (1990) study, low-ranking female vervets may have been forced to feed at feeding sites that were less productive and, thus, had to spend more time moving, or they may have had to interrupt their feeding more often than high-ranking females. Either of these possibilities would account for the findings that low-ranking females fed at more sites per tree, at more sites per patch, and at more trees per unit time. However, because low-ranking females fed longer on defended foods and more slowly in general than high-ranking females, they may not have been as efficient feeders as high-ranking females or had access to lower-quality feeding sites. In the next chapter, I explore further the feeding competition and dominance relations of adult female vervets, specifically comparing their behavior within two different habitats.

# 5

# Food Availability's Influence on Competition and Dominance in Vervets

Because food availability is expected to be most limiting to an adult female's reproductive success relative to other factors, such as availability of mates (Trivers 1972; Wrangham 1980), I examined contests over foods and resulting dominance relationships in detail. I was able to study contest competition and dominance among adult female vervets on Segera as they used two distinct types of habitat that exhibited differing patterns of food availability. Whereas a number of studies have demonstrated the effects that food availability has on agonism over foods between individuals within a group (Cheney & Seyfarth 1990; Izar 2004; Roeder & Fornasieri 1995; Struhsaker & Gartlan 1970; Symington 1988; Whitten 1983; Wrangham 1981), it is less clear that competition over foods produces the dominance patterns predicted by the models of female social behavior. I specifically investigate the assumed link between contest competition over food and the resulting female dominance relationships for vervets on Segera.

Contest competition is hypothesized to lead to rank-related differences in feeding behavior that ultimately result in varying reproductive success among adult females of contrasting rank. Priority of access to resources and increased foraging efficiency are often used as indirect measures of reproductive success. For example, Wrangham (1981) found that high-ranking female vervets successfully outcompeted low-ranking females over access to water sources during times of water shortage, and low-ranking females showed significantly higher rates of mortality during this period ($N = 3$ of four deaths in a group with eight adult females). Harcourt (1987) summarized data from twenty studies of nine species of free-ranging, provisioned, and captive nonhuman primates regarding the effects of dominance on adult females' reproductive success. He concluded that, for Old World monkeys, a correlation between "stress" from competition and fertility has yet to be demonstrated, but that there is evidence that fertility could be affected by competition's effect on nutrition. He also noted that dominant animals were likely to produce more offspring than subordinates when foods were clumped (Harcourt 1987). Whitten (1983) demonstrated such an effect for vervets in Samburu, Kenya. High-ranking female vervets had lower rates of infant mortality than low-ranking females when their main foods (e.g., flowers of *Acacia tortilis*) were clumped. There appears to be a link, in many cases, between feeding competition and dominance; however, this link has not yet been examined closely for many species.

## THE CONCEPT OF DOMINANCE

The term *dominance* has been much used and carries many connotations (Drews 1993; see Bernstein 1981). Here, "dominance" is used as a structural definition to describe an observed pattern of behavior (after Drews 1993). Specifically, the dominance "relationships" of individuals are examined. Relationships "involve a series of interactions between two individuals known to each other, and can be characterized by the content, quality, and patterning of those interactions" (Hinde 1979: 296). Interactions involve particular behaviors and the quality of those behaviors (Hinde 1979). "Agonistic interactions" are used here to describe interactions between individuals where a winner of any single dyadic encounter is the dominant individual *for that interaction* and the loser of this encounter is the subordinate individual *for that interaction* (modified from Drews 1993). Agonistic interactions are representative of contest competition between individuals and include such behaviors as supplants and approach-retreat interactions. Consequently, dominance is used to describe the position of one individual relative to others within a hierarchy (i.e., the relationships of the individuals studied as assessed through their behavioral interactions). Within this hierarchy, more dominant individuals consistently win in dyadic interactions with particular individuals, and subordinate individuals consistently lose in dyadic interactions with particular individuals. A win or a loss is defined as access to resources. Lockwood (1979) terms this the tournament model of dominance.

Of the twelve common definitions of dominance reviewed by Drews (1993), I use the original description of "peck-order" by Schjelderupp-Ebbe (1922), modified to include more than one type of agonistic interaction (Barrette & Vandal 1986). According to this definition, individual B always or usually submits to individual A; individual C always or usually submits to individual B and to individual A; and so on (A > B > C and A > C). It is implicit that individuals recognize one another, and a dominant individual will rarely have to resort to physical aggression to overcome a subordinate (Barrette & Vandal 1986; Schjelderupp-Ebbe 1922). The result is a consistent unidirectional pattern that occurs over time (Schjelderupp-Ebbe 1922). Such a pattern indicates that particular members of the social group have priority of access to resources in competitive situations (Clutton-Brock & Harvey 1976) and/or receive significantly less aggression from subordinates. The assumption, according to ecological and socioecological models, is that subordinates ultimately suffer fitness costs relative to more dominant individuals.

Dominance relationships may differ according to the specific resource that is limiting to individuals. Popp and DeVore (1979) note, "dominance hierarchies are expected to be time-and-resource-specific" (Popp & DeVore 1979: 331). This expectation is based on several assumptions according to a model of optimal competitive strategies: (1) resources may differ in value for an individual, (2) resources may differ in value to an individual over time, and (3) the value of resources may differ for individuals (Popp & DeVore 1979).

## TESTING HYPOTHESES

Female social relationships are predicted to vary according to key factors such as food availability (Isbell 1991; van Schaik 1989; Wrangham 1980). Females are expected to overtly compete for food resources when those food resources are defendable (e.g., when foods are clumped or otherwise worth usurping). Aggressive, or contest, competition is expected to lead to dominance hierarchies within social groups (Isbell 1991; van Schaik 1989; Wrangham 1980). Because contests over foods by adult female primates are expected to lead ultimately to differential reproductive success under a very specific set of conditions (Wrangham 1980), I compared food-related competition and dominance with that in other contexts. On the basis of the models, I predicted that adult female vervets would exhibit different patterns of contest competition (as measured by agonistic interactions) in habitats with contrasting patterns of food availability. I also predicted that agonism among vervets would occur more frequently in riverine than in whistling-thorn woodland because foods were more clumped in the former habitat.

Agonistic interactions that occur during feeding (versus those that occur in nonfeeding contexts) and resultant dominance relationships are expected to reflect the importance of food availability to females' reproduction and survival. If food is limited, food-related contests are expected to be more frequent and intense than other contests (Shopland 1987). Adult female dominance relationships that arise as a result of contest competition among females are predicted to vary with food availability as well, although to a lesser degree than are agonistic interactions. This is due to the fact that individuals contest a wide variety of resources, not just food. It is possible that a dominance hierarchy may be evident in contexts such as feeding, but absent or undetectable in other contexts. For example, studies of agonism among brown capuchins (*Cebus apella*) found that most aggression occurred at fruit trees, and that the rate of agonism during feeding was higher than during nonfeeding contexts (Janson 1985). Additionally, the rate of agonism differed significantly according to food type during the dry season (Janson 1985). Adult female dominance hierarchies among vervets are consistently reported to be linear and stable (Cheney & Seyfarth 1990; Whitten 1983; Wrangham 1981). However, in some primate species, such as langurs (*P. entellus*), the strength of the dominance hierarchy varies with the abundance and distribution of foods (Koenig et al. 1998). I test the hypothesis that the strength of the dominance hierarchy among adult female vervets also varies (as does contest competition) under differing conditions of food availability, such as that characteristic of their two main habitats.

Systematic data on agonistic interactions were collected during 37,089 contact minutes with vervets. An attempt was made to follow each vervet study group from approximately 0730 to 1600 hours, for four days per month. From July 1993 to December 1994, an average of 33.2 h per month was spent in contact with the main vervet study group, i.e., Pond group. From January to March 1995, the Pond study group was observed on average 6.6 h per month. Data were recorded on (1) the identity of individuals involved in agonism, (2) intensity of agonistic interactions, (3) habitat, (4) time of day, and (5) context of interaction, such as whether it occurred over food, grooming partners, mating partners, or space. Agonistic interactions included any approach or chase by one individual and the subsequent retreat of another individual (Table 5–1). More subtle agonistic interactions included cringing, crouching, or shrinking away (after Hall 1967) from an approaching individual who was not necessarily directing attention at the individual being approached, being mounted by another individual, and interrupting activity at the approach of another individual (see Table 5–1). I constructed dominance matrices using all agonistic interactions recorded during the course of the study, which were collected during focal animal follows and *ad libitum*. The outcomes of dyadic contests are scored within cells of the matrix and indicate winners (row scores) and losers (column scores) in agonistic contests. Species that exhibit stable, linear dominance hierarchies should exhibit few scores that fall below the diagonal of the matrix (which would indicate reversals against the hierarchy) and few empty cells (which indicate unknown or undecided relationships).

**Table 5–1** Behavioral Catalog of Agonistic Interactions

| Behavior | Operational Definition | Intensity Score* |
|---|---|---|
| Pause | One individual stops action and focuses attention on approaching individual; after approaching individual has passed, subject resumes action | 1 |
| Mount | Ventral surface of one individual comes into contact with dorsal surface of another; mounted individual may or may not resume behavior | 2 |
| Cringe | Includes crouch, flinch, lean away from approaching individual. Monkey lowers its body by flexing its arms and legs; may tuck its head between drawn-up shoulders; After van Hoof (1974) | 3 |
| Avoid | One individual moves away from approaching individual to a distance of no more than 2 m | 4 |
| Leave | One individual moves away from an approaching individual to a distance of 2 m or more | 5 |
| Supplant | One individual displaces another individual from a specific location and exhibits the same activity as the individual that was displaced | 6 |
| Chase-flee | One individual runs after another individual, who flees. Play behavior not included | 7 |
| Physical aggression | Includes biting, slapping, pulling, and other bodily contact indicative of fighting between two individuals | 8 |

*Most intense interaction type = 8, least intense interaction type = 1.

## RESULTS: CONTEST COMPETITION AND DOMINANCE IN VERVETS

The majority of agonistic interactions observed among vervets occurred between adult females (46%, $N = 219$) (Pruetz 1999). Agonistic interactions in which at least one participant was an adult female comprised 77% of all observed interactions ($N = 171$). Adult females interacted agonistically with adult males (19% of female agonism, $N = 32$) as often as with immature vervets (19% of female agonism, $N = 32$) but most often with each other (63% of female agonism, $N = 107$).

When context of the interaction could be identified, the majority of agonistic interactions between adult female vervets occurred during feeding (73.8%, $N = 79$) (Table 5–2). However, this occurred at a rate of only 0.10 contests per feeding/foraging hour. A total of thirty-one contests could not be relegated to a single context. Other contexts in which agonism occurred between females included competition over space ($N = 4$), infants ($N = 8$), grooming partners ($N = 7$), and in support of an individual ($N = 3$). Agonism occurring in the vicinity of a termite mound was probably food

**Table 5–2** Context of Agonistic Interactions Involving Adult Female Vervets

| Context | Female–Female | Female–Male | Female–Immature | Total |
|---|---|---|---|---|
| While feeding | 79 | 23 | 24 | 126 |
| Grooming partner | 7 | 0 | 1 | 8 |
| While resting | 4 | 1 | 0 | 5 |
| Proximity to an individual | 1 | 0 | 0 | 1 |
| In support (coalition) | 3 | 2 | 2 | 7 |
| Infant access | 8 | 0 | 0 | 8 |
| Seat | 4 | 1 | 0 | 5 |
| Retaliation | 0 | 2 | 1 | 3 |
| Redirection | 0 | 2 | 0 | 2 |
| Termite mound | 1 | 1 | 4 | 6 |
| **Total** | **107** | **32** | **32** | **171** |

related but could not be absolutely identified as such by the observer ($N = 1$). Each of the other contexts besides feeding accounted for less than 8% of interactions observed between females ($N = 27$; see Table 5–2).

The intensity of agonistic interactions among adult females was also analyzed (Table 5–3). Fighting was considered the most intense interaction type, whereas "pause" was considered the least intense. Approach-avoid was the most common agonistic interaction among adult females (37% of contests, $N = 66$). Supplants accounted for 18% of all agonistic interactions observed ($N = 25$). Adult female vervets engaged in little physical aggression (0.7% of contests, $N = 1$). The highest proportion of physical fights occurred between adult males and females (44%, total $N = 9$), whereas the highest proportion of chase-flee interactions occurred between adult female and immature vervets (32%, total $N = 15$).

**Table 5–3** Intensity of Agonistic Interactions Exhibited by Adult Female Vervets

| Type | Adult Females | Female–Male | Female–Immature | Row Totals |
|---|---|---|---|---|
| Fight | 1 | 4 | 2 | 7 |
| Chase-flee | 9 | 9 | 15 | 33 |
| Supplant | 25 | 8 | 13 | 46 |
| Leave | 28 | 6 | 0 | 34 |
| Avoid | 66 | 14 | 10 | 90 |
| Cringe | 4 | 10 | 1 | 15 |
| Mount | 4 | 0 | 0 | 4 |
| Pause | 1 | 0 | 0 | 1 |
| **Total** | **138** | **51** | **41** | **230** |

Because contest competition during feeding was hypothesized to indicate the degree to which food availability limited adult female vervets, the intensity of agonism between adult females during feeding ($N = 79$ contests) versus nonfeeding contexts ($N = 59$ contests) was further explored (Figure 5–1). A larger proportion of chase-flee interactions occurred during nonfeeding contests ($N = 7$) than during feeding contests ($N = 2$). The proportion of supplants ($N = 16$) and approach-avoid interactions ($N = 43$) was higher in feeding compared to nonfeeding contexts ($N = 9$ and $N = 23$, respectively) whereas the proportion of approach-leave interactions was higher during nonfeeding contexts ($N = 16$) than during feeding contexts ($N = 9$). A two-way Chi-square test showed that feeding versus nonfeeding contests differed significantly in their intensity ($\chi^2 = 15.9$, df $= 7$, $p < 0.05$). Contests that occurred during nonfeeding were more intense than those during feeding.

A total of thirty-seven agonistic interactions for which context could be recorded occurred between adult females in whistling-thorn woodland. In riverine habitat, eighteen agonistic interactions were observed where context could be scored. Different amounts of time were spent feeding in the two habitats by vervets, and the rates of agonistic interactions were adjusted to account for these differences. Using the amount of time adult females were observed in the different habitats (69% whistling thorn, 31% fever tree), and the amount of time spent foraging by adult females in these habitats from September to December 1994 (76% whistling thorn,

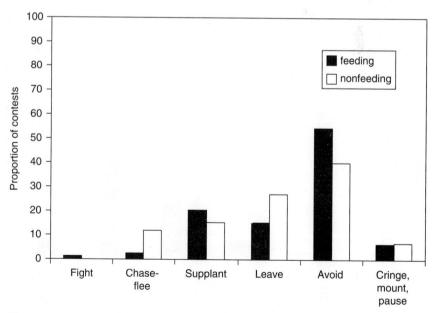

**Figure 5–1**  Intensity of adult female feeding contests.

**Table 5–4** Agonistic Interactions in Female Vervets in Different Habitats

| Context | Whistling-Thorn Woodland | Fever Tree Riverine | Chi-Square Tests |
|---|:---:|:---:|---|
| Observed food related | N = 37 | N = 18 | — |
| Weighted food related | 0.11 per hour | 0.30 per hour | $\chi^2 = 32.82$, df = 1, $p < 0.001$ |
| Observed nonfood related | N = 5 | N = 13 | — |
| Weighted nonfood related | 0.05 per hour | 0.10 per hour | $\chi^2 = 5.00$, df = 1, $p < 0.05$ |

31% riverine), the number of agonistic interactions was presented as a weighted rate per hour within the respective habitats (Table 5–4). The weighted rate of food-related agonism was significantly lower in whistling-thorn woodland compared to riverine habitat. More than twice as many interactions occurred per feeding hour in the riverine habitat compared to whistling-thorn woodland.

The intensity of agonistic interactions in the two different habitats also was examined (Figure 5–2). Avoiding an approaching individual accounted for the majority (51%, $N = 44$) of all agonistic interactions among adult females in each of the habitats.

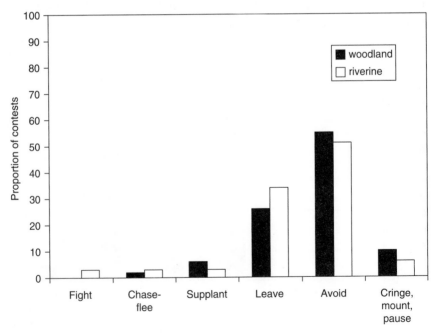

**Figure 5–2** Intensity of agonism among female vervets in different habitats.

**Table 5–5**  Adult Female Vervet Dominance Matrix

|      | crv | frj | chi | sal | tor | moo | qso | bur |
|------|-----|-----|-----|-----|-----|-----|-----|-----|
| crv  |     | 6   | 9   | 8   | 2   | 5   |     | 3   |
| frj  |     |     | 22  | 10  | 4   | 3   |     | 6   |
| chi  |     |     |     | 9   | 9   | 8   | 2   | 8   |
| sal  |     |     |     |     | 4   | 5   | 1   | 5   |
| tor  |     |     |     |     |     | 4   | 2   | 3   |
| moo  |     | 1   |     |     |     |     | 1   | 2   |
| qso  |     |     |     |     |     |     |     | 9   |
| bur  |     |     |     |     |     |     |     |     |

I used a dominance matrix to illustrate the dominance relationships of adult female vervets in the Pond study group. Scores in columns indicate the number of times a female was subordinate during an agonistic interaction, whereas those in rows indicate the number of times she was dominant. The dominance hierarchy observed for adult female vervets was linear and stable (Table 5–5) with only two reversals (less than 2% of interactions observed) and two unknown relationships. One of the reversals was between the only female that matured during the study (MOO) and a higher-ranking adult female. Appleby's (1983) adaptation of Kendall's (1962) method to test for significance of linearity was used to determine both the degree of linearity of the hierarchy and the number of circular triads within the hierarchy. The number of circular triads is denoted by the $d$-value [using the formula: $d = (N(N-1)(2N-1)/12) - (1/2)$(number of subordinate individuals for each subject[2], Appleby 1983]. On the basis of these findings, the proportion of triads in which relationships are linear as opposed to circular is greater than that expected by chance (Appleby 1983). The coefficient $K$ denotes values from 0 to 1, with a value of 0 indicating complete absence of linearity and a value of 1 indicating a linear hierarchy [using the formula: $K = (1 - 24d)/N^3 - 4N$ for an even $N$; Appleby 1983]. This method controls for any reversals or unknown relationships that could be inconsistent with a linear ranking as well as taking into account linearity due to chance (Appleby 1983). One adult female (MND) was omitted from calculations of linearity. She was observed in only two agonistic interactions and disappeared in May 1994. The dominance hierarchy for adult female vervets was transitive and thus genuinely linear ($d = 2$, $K = 0.90$, $p < 0.006$).

The mean number of agonistic interactions observed for individual adult female vervets other than MND during the study period was 37.8 (range 15–67). Of the eight Pond group females, the four highest-ranking females directed significantly more agonism toward others (mean 32.3) than did the four lowest-ranking females (mean 5.5) ($t = 4.018$, df $= 3.7$,

**Table 5–6**  Adult Female Vervet Dominance Matrix: Whistling-Thorn Habitat

|     | crv | frj | chl | sal | tor | moo | qso | tor |
| --- | --- | --- | --- | --- | --- | --- | --- | --- |
| crv |     | 3   | 1   | 2   |     | 3   |     | 1   |
| frj |     |     | 7   | 2   | 1   |     |     | 3   |
| chl |     |     |     | 3   | 5   | 2   |     | 1   |
| sal |     |     |     |     | 2   | 2   |     | 2   |
| tor |     |     |     |     |     |     | 1   | 1   |
| moo |     |     |     |     |     |     |     |     |
| qso |     |     |     |     |     |     |     | 2   |
| bur |     |     |     |     |     |     |     |     |

$p = 0.018$, one-tailed). Higher-ranking female vervets (mean 16.3) did not receive significantly less agonism than did lower-ranking females, however (mean 21.5) ($t = -0.536$, df = 5.8, $p = 0.612$, one-tailed).

Agonistic interactions were categorized according to whether they occurred in the riverine habitat ($N = 113$ interactions) or whistling-thorn woodland ($N = 44$ interactions), and dominance matrices were constructed for each habitat (Tables 5–6 and 5–7). The one reversal among female vervets occurred in the riverine habitat. The dominance hierarchy for adult females in whistling-thorn woodland included a greater proportion of unknown relationships ($N = 9$ or 32% when MND is excluded) than the hierarchy constructed for contests that occurred within riverine habitat ($N = 2$ or 7%). When the hierarchy is analyzed statistically for whistling-thorn woodland, the linearity is not significant ($d = 10.5$, $K = 0.475$, $p > 0.15$) even though no reversals were recorded. The nonlinearity is likely due to the high number of dyads ($N = 9$) for which no agonistic interactions occurred. The degree of linearity of the hierarchy for adult female vervets when they used the riverine habitat is similar to the overall hierarchy, being linear ($K = 0.90$), with significantly fewer circular triads than expected ($d = 2$, df = 21, $p < 0.006$).

**Table 5–7**  Adult Female Vervet Dominance Matrix: Riverine Habitat

|     | crv | frj | chi | sal | tor | moo | qso | bur |
| --- | --- | --- | --- | --- | --- | --- | --- | --- |
| crv |     | 3   | 8   | 6   | 2   | 2   |     | 2   |
| frj |     |     | 15  | 10  | 3   | 3   |     | 3   |
| chi |     |     |     | 6   | 4   | 6   | 2   | 7   |
| sal |     |     |     |     | 2   | 3   | 1   | 3   |
| tor |     |     |     |     |     | 4   | 1   | 2   |
| moo |     | 1   |     |     |     |     | 1   | 2   |
| qso |     |     |     |     |     |     |     | 7   |
| bur |     |     |     |     |     |     |     |     |

**Table 5–8**  Adult Female Vervets' Dominance Matrix: Food-Related Contests

|     | crv | frj | chl | sal | tor | moo | qso | bur |
|-----|-----|-----|-----|-----|-----|-----|-----|-----|
| crv |     | 5   | 6   | 5   | 2   | 3   |     | 3   |
| frj |     |     | 10  | 5   | 1   | 2   |     | 3   |
| chl |     |     |     | 5   | 5   | 5   |     | 1   |
| sal |     |     |     |     | 3   | 2   |     | 3   |
| tor |     |     |     |     |     |     | 1   |     |
| moo |     |     |     |     |     |     |     | 1   |
| qso |     |     |     |     |     |     |     | 2   |
| bur |     |     |     |     |     |     |     |     |

The possibility that food-related contest competition accounted for the linearity of the adult female vervets' hierarchy also was explored (Table 5–8). When a dominance matrix was constructed using agonistic interactions over food only, the degree of linearity was high ($K = 0.89$), and the number of circular triads was significantly less than expected ($d = 2.25$, df $= 21, p < 0.006$). A matrix that was constructed using contests over food in the riverine habitat did not produce a statistically linear hierarchy, however (see Pruetz 1999). The degree of linearity was low ($K = 0.120$), and the proportion of circular triads was not significantly less than expected ($d = 16.75, p > 0.153$).

Matrices were constructed for food-related contests that occurred during the dry and wet seasons in order to explore whether dry season scarcity of foods might explain the linearity of female vervets' hierarchy (Pruetz 1999). However, neither of these hierarchies exhibited a high degree of linearity, nor were the proportions of circular triads significantly less than expected. The proportion of food-related contests that occurred during dry season months was similar to the proportion of these contests that occurred during wet season months, although the rate of contests per hour and the rate of food-related contests per hour were higher in the wet season (Pruetz 1999).

## DOMINANCE PATTERNS AND FEEDING COMPETITION IN VERVETS

I predicted that rates of agonism would reflect the differences in patterns of food availability within the two habitats vervets used. As predicted, the rate of agonism shown by female vervets during foraging was significantly lower (0.11 interactions per hour, $\chi^2 = 32.82, p < 0.001$, df $= 1$) in whistling-thorn woodland where foods are less clumped than in riverine habitat (0.30 interactions per hour). Because the outcome of food-related contests may be more costly to adult females in terms of resources lost than for contests over space or grooming partners, contests over foods were predicted to be more frequent and intense than contests over non-foods. The results, however, are unclear. During agonism related to feeding, the proportion of supplants and approach-avoid interactions was higher than the proportion occurring during nonfeeding agonism, but the proportion of leave-approach interactions was higher in nonfeeding contexts. The proportion of chase-flee interactions was higher for nonfood contexts. This is not surprising considering the contested object is likely to be lost if a dominant individual pursues the subordinate.

In Chapter 3, I demonstrated that foods within the two different habitats used by vervets differed in both abundance and distribution. Fever trees are the vervets' main food source in the riverine habitat, whereas whistling thorn is the main food source within the woodland habitat (Pruetz & Isbell 2000). In the riverine habitat, fever trees were dispersed, on average, every 13.3 m and occurred at a mean density of 57 trees per hectare. In contrast, in the woodland habitat, whistling-thorn trees were dispersed every 2.4 m on average and occurred at a mean density of 2,477 trees per hectare. The distribution of whistling-thorn trees within the vervets' home range was also random. In comparison with the fever tree *Acacia*, whistling thorn can be described as superabundant for vervets. Based on the proportion of the vervets' home range that consisted of riverine (11.2 ha) versus woodland (28.8 ha), vervets had access to approximately 635 fever trees and 71,337 whistling-thorn trees of all sizes within their home range.

Although absolute counts of food items other than gums were not recorded for fever trees, I estimated the abundance of foods within whistling-thorn versus fever tree *Acacia*s using data on crown size. Average crown volume per individual fever tree was 65.4 m$^3$ (Pruetz 1999) and thus, approximately 3,728 m$^3$ of fever tree crown (i.e., foraging area) was available to vervets per hectare within the riverine habitat, for a total of 41,751 m$^3$ of fever tree *Acacia* crown available to vervets within their home range. Depending on tree height, average crown volume of whistling-thorn trees ranged from 0.1 to 4.5 m$^3$ (Pruetz 1999). Because the availability of trees of different heights was measured, the average crown volume of whistling thorn could be calculated by weighting the values of crown volume for the different height categories by the availability of those trees within the home range of vervets. Average crown volume of whistling-thorn trees per hectare within the woodland portion of the vervets' home range was thus calculated as 2,379 m$^3$. The estimated crown volume of whistling-thorn trees within the vervets' home range was 68,503 m$^3$. Therefore, although crown volume was less for whistling thorn per hectare, the overall volume of whistling-thorn trees available to vervets was greater than that of fever tree *Acacia*.

Vervets did contest whistling-thorn foods, however. Vervets supplanted one another over whistling-thorn swollen thorns, although rarely, contrary to my expectations. This food was rated lowest in contestability (see Table 3–8). No other whistling-thorn food items were contested in supplants between vervets, but less severe agonism over whistling-thorn new growth, seeds, and gum did occur. Vervets contested new growth more than gum and seeds, but I initially ranked new growth lower in contestability. New growth was ranked higher than seeds in processing costs, higher than gum in nutritive value, and lower than gum or seeds in distribution. Seeds were less abundant than any other whistling-thorn food item other than swollen thorns and were ranked lower than all other food items other than swollen thorns in distribution.

Whereas the overall pattern of female contest competition during feeding within the two habitats used by vervets corresponded with the observed patterns of food availability as predicted by the ecological models, other variables had to be considered when attempting to understand female feeding competition. For example, even foods within the whistling-thorn woodland were contested, although at a lower rate than contests in the riverine habitat. Contests over whistling-thorn foods accounted for 29% of all female vervets' contests over identified foods in the whistling-thorn woodland. In the next chapter, I specifically investigate the various properties of foods in terms of their influence on contest competition in vervets and patas monkeys.

Whereas van Schaik (1989) emphasizes the importance of food distribution in predicting social relations among female primates living in the same social group, and Wrangham (1980) and Isbell (1991) stress the importance of food abundance, I consider both of these aspects of food availability to be important influences on female primate competitive relationships within groups. Variables such as nutritional quality and costs related to processing foods also should be considered. A temporal measure of food availability—food site depletion time (Isbell et al. 1998a,b)—considers food abundance but does not take into account food nutritional quality or the spacing patterns of food sites. A more accurate term in this study might be "food site departure time," because monkeys are not depleting all available whistling-thorn foods. Ant protection caused primates to move to other patches after eating only a few items. Initially, I defined availability according to the abundance and distribution of foods. Upon examination of the results of this study, I would expand this definition to include the nutritional quality of foods, as well as their processing costs.

Ecological factors played an important role in influencing within-group agonism, and, thus, dominance relationships among individuals in the vervets studied here. Rates of agonistic interactions were low but varied according to habitat, and food availability correlated positively with these records of increased agonism. Habitat with more locally abundant foods was associated with higher rates of agonism (0.40 in riverine versus 0.16

in whistling-thorn woodland). However, whether these rates are significant in affecting adult females' feeding behavior and, ultimately, reproductive success is in no way clear. The availability of fever trees compared to whistling-thorn trees and other foods within the two habitats used by vervets on Segera suggests that contest competition is rare but more prevalent when foods are more clumped in space. However, within the riverine habitat, food-related contests accounted for just over half (58%, $N = 18$) of all contests. Contests over grooming partners, infants, and during resting accounted for the remaining contests (42%, $N = 13$). Food-related agonism produced a significantly linear hierarchy among female vervets, but agonism that occurred in both habitats was needed to produce this result. Contests over foods that occurred in the riverine habitat, where foods were more "patchily" distributed (i.e., in larger clumps more widely separated in space), could not by themselves account for the linearity of females' hierarchy. A significantly linear hierarchy was only apparent if all contests (food related and nonfood related) within the riverine habitat were considered or if all food-related contests (those that occurred within both the habitats) were considered. The predictions of models of female social relationships among primates (van Schaik 1989; Wrangham 1980), in this case, are not supported. The dominance hierarchy exhibited by adult female vervets when using the woodland habitat was not significantly linear. However, the pattern of linearity did not deviate from the overall hierarchy observed among female vervets, i.e., there were no reversals. The absence of decided contests among all dyads during feeding in whistling-thorn woodland indicates that feeding competition (but not necessarily dominance) is not particularly important to vervets in this habitat. Such a finding coincides with results indicating that foods within the woodland habitat are randomly scattered. An aspect of behavior that was not investigated for monkeys in this study was the role of dominance and individual spatial position within a group. Ron and co-workers (1996) found that, among chacma baboons, more subordinate individuals occupied spatial positions at the periphery of the troop. These positions were interpreted as being less safe, in terms of predation pressure, compared to the central positions occupied by more dominant females (Ron, Henzi, & Motro 1996). This would be an additional variable to consider in terms of patas monkey and vervet behavior in the predator-rich environment at Segera.

Although vervets on Segera used a habitat where foods were relatively clumped and abundant in space, they foraged and fed more often in a habitat where foods were randomly distributed in relatively small patches and were abundant (i.e., whistling-thorn woodland). Based on the amount of time adult female vervets spent feeding and foraging in the different habitats and the time spent there, females spend an estimated 4.2 h foraging/feeding in woodland and 0.77 h foraging/feeding in the riverine habitat during an 8-h day (time period during which vervets were observed in this study). Even if it is assumed that the vervets in this study use the riverine habitat for an additional 4 h per day when they were not observed (i.e., 2 h before and after sampling periods), females would spend a total of only 2 h feeding/foraging here per day. Vervets on Segera spend over twice as much time feeding/foraging within the woodland habitat where foods are abundant and scattered, yet the linear dominance hierarchy of females, according to the models, is indicative of clumped foods. This contradicts the expectations of the socioecological models.

Food-related agonism does explain the linearity of the female vervets' hierarchy. In this case, relying on patterns of food availability to predict female social relationships within a group was not sufficient. Clearly, reevaluation of current models used to predict female social relationships among primates is necessary based on these data.

## THE SIGNIFICANCE OF DOMINANCE TO VERVETS

Stable, linear dominance hierarchies are thought to be indicative of variable resources and variation in individuals' access to these resources (Wrangham 1980). Among female primates, dominance is thought to provide priority of access to resources that consequently results in rank-related differences in reproductive success. Ultimately, in order to assess the influence of feeding competition on adult female primates' social behavior, data on the reproductive success of females of differing ranks are necessary. Although estimates of reproductive success are lacking in this study, data are available on the ecological costs and benefits of rank for vervets on Segera. Rank-related differences in behavior contributing to reproductive success, such as feeding behavior under conditions when resources are limited, can be viewed as indirect indicators of rank-related differences in reproductive success. High-ranking adult female vervets in this study did not feed as long per bout on defended food items of whistling-thorn *Acacia;* spent more time foraging; fed at fewer whistling-thorn feeding trees, feeding sites, and sites per tree per unit time; and fed at fewer feeding sites of fever tree *Acacia* per unit time than low-ranking females. Low-ranking female vervets may have spent more time traveling than foraging, but alternatively, they could have been selecting foods that were more quickly processed. This is unlikely, however, because low-ranking females fed at more trees and feeding sites per tree on each of the major *Acacia* tree species. Low-ranking females may be forced to spend less time per feeding site than high-ranking females, by having to feed in patches with less food available or having to evacuate food sites more often. However, the dimensions of patches used by high- and low-ranking female vervets did not differ (Pruetz 1999). Additionally, there was no significant difference in processing times of defended whistling-thorn foods according to rank, although high-ranking females did tend to feed faster. Thus it appears that, for vervet females on Segera, whereas rank-related differences in feeding did occur, the importance of these differences to females' feeding efficiency or other behaviors influencing reproductive success is unclear.

## DOMINANCE STYLE IN SEGERA VERVETS

Vervets have been classified, along with baboons and macaques, as species that exhibit stable, linear, and nepotistic dominance hierarchies (Gouzoules & Gouzoules 1987; Isbell & Pruetz 2000; Kaplan 1987; Melnick & Pearl 1987; Seyfarth 1980; Walters & Seyfarth 1987). The type of female social relationships characterized by such a hierarchy (i.e., despotic, nepotistic) has been linked to the patterns of foods characteristic of vervets' ranges (i.e., clumped). Such a pattern of food availability is thought to promote strong contest competition (van Schaik 1989) and to consequently result in the characteristic "cercopithecine dominance style." The data supporting this conclusion, however, originate from only two sites in East Africa—Amboseli and Samburu, Kenya—and were not supported in the current study.

The Amboseli vervet population was studied by T. Struhsaker in the mid-1960s, by D. Klein in 1975, and by D. Cheney, R. Seyfarth, and a number of their colleagues from 1977 to 1988. The population declined in numbers from the mid-1960s to 1988 (Cheney & Seyfarth 1990). In 1977, the study population consisted of 215 individuals in eleven groups, but, in 1988, only 35 individuals in four groups remained (Cheney & Seyfarth 1990). Habitat destruction was the ultimate cause of this decline, as the water table in Amboseli rose and caused a die-off of fever tree *Acacias*. A population of elephants that were, for the most part, restricted to the national park area also contributed to the decline in the numbers of *Acacia* trees. These factors brought about an increase in mortality due to predation (Cheney & Seyfarth 1990; Isbell, Cheney, & Seyfarth 1990). Thus, behavior characterizing the Amboseli vervet population might be a response to an extreme environmental situation. Each of the three vervet groups studied by Seyfarth (1980) from 1977 to 1978 exhibited significantly linear dominance hierarchies ($d = 0$, $p < 0.001$, $K = 1.0$ for all groups, my calculations), with reversals against the hierarchy accounting for 4%, 0.5%, and 0.1% of approach-retreat interactions. A stable and linear dominance hierarchy in an environment where resources are so limited is not unexpected but does not necessarily present a typical view of vervets' social interactions. How limited were resources at Amboseli during 1977–1978, however? This thesis stresses the importance of quantifying food availability in ways that are comparable and replicable between sites. Rainfall data are extremely imprecise measures of food availability (Glander 1978), yet are the only comparable data available for Amboseli and Segera. Amboseli receives <300 mm of rainfall per year, compared to approximately 700 mm at the Segera site. This crude measure indicates, at the very least, that vegetative biomass at the Amboseli site is less than that at the Segera site, even when discounting the habitat destruction characteristic of the Amboseli study site.

The vervets studied at Samburu, Kenya, are often used as examples of the characteristic female nepotistic, despotic dominance hierarchy (Fedigan 1993; Horrocks & Hunte 1993; Melnick & Pearl 1987), but analyses of Whitten's (1983) data using Appleby's (1983) method of controlling for the number of circular relationships within a group produce a statistically nonlinear hierarchy. In each group of vervets studied ($N = 2$), the degree of linearity was low ($K = 0.306$, $K = 0.334$), and hierarchies were not significantly linear ($d = 27.8, p > 0.131$ and $d = 36.6, p > 0.20$, respectively, my calculations). Although the dominance hierarchies in Whitten's groups were not significantly linear, high-ranking females exhibited a higher birth rate when preferred food species (i.e., flowers of *Acacia tortilis*) were clumped in distribution (Whitten 1983). Whitten constructed her dominance matrices using supplants over food and/or space only, whereas the current study and many others have used data on all agonistic behavior. Because Whitten also found rank-related differences in reproductive success, should we assume that supplant rates over food and space did not accurately represent all of the interactions among females that were relevant to their reproductive success? Watts (1994) found that female mountain gorillas could be linearly ranked using approach-retreat interactions but not using more intense forms of agonism alone. If models such as van Schaik's (1989) are to predict female dominance relationships using competitive behaviors, what specific agonistic behaviors are relevant?

## "TYPICAL" CERCOPITHECINES?

Whitten's (1983) findings on reproductive success indicate that rank-related differences are biologically important to female vervets under certain circumstances (i.e., when foods are clumped), even though her measures of dominance relations failed to indicate as much (i.e., hierarchies were not significantly linear). Accordingly, vervets should not be classified along with baboons and macaques as primate species in which contest competition is "strong" (van Schaik 1989) because their hierarchy was not linear. The rate of agonistic contests is comparable between Whitten's vervet female study subjects, baboons, and macaques, however.

Vervet females on Segera also differ from previously studied populations and from "typical Cercopithecines" such as many baboon and macaque species in that they exhibit relatively little alliance or support behavior. Only 3% of interactions between adult female vervets involved the support of an ally to one of the individuals involved in the contest. Of all contests in which adult females were participants, only 4% entailed support of one individual by another.

## WHY A STABLE, LINEAR DOMINANCE HIERARCHY ON SEGERA?

Evidence presented in this study suggests that feeding contest competition is relatively unimportant for the adult female vervets on Segera in comparison with vervets at other sites and other primate species such as baboons and macaques. Yet, female vervets on Segera form stable, linear dominance hierarchies, and contests over foods produce a statistically linear hierarchy. Considering the "dominance style" of vervets in relation to other guenons as well as macaques and baboons may help determine what factors induce vervet females to form despotic social relationships.

Vervets differ from patas monkeys and other guenons in that females exhibit stable and linear dominance hierarchies. Vervets also differ from other guenons and patas monkeys in that they exhibit a multi-male social grouping system. Perhaps the inclusion of multiple males in a social group produces increased competition between adult females. Female vervets compete with males for food (14% of all females' contests, $N = 23$), and individual females are subordinate to individual males (Cheney & Seyfarth 1990). More intense competition with adult males may induce female vervets to more actively compete with one another compared to females that live in single male social groups. Although an alpha female vervet is dominant to all other females in her social group, she is subordinate to all males in her social group. In the case of the Segera study group, the alpha female (CRV) was dominant to six to seven adult females but subordinate to six to eight adult males during the study period. However, free-ranging vervets in a historically introduced population on Barbados do form one-male social groups (Horrocks & Hunte 1993) and also exhibit statistically linear female dominance hierarchies (Horrocks & Hunte 1983). Barbados vervets spend a large proportion of feeding time on cultivated foods, which may impact feeding competition and social relationships. For example, although patas monkeys in the current study did not exhibit stable, linear dominance hierarchies, a provisioned population of patas monkeys in Puerto Rico was reported to exhibit linear dominance hierarchies (Zucker 1994). The utility of the Barbados population for understanding the behavior of naturally occurring vervets is therefore limited.

Studies of agonism and dominance among female primates assume that dominance hierarchies reflect patterns of food availability. This is an example of circular reasoning. For example, Koenig and co-workers (1998) supported their conclusions that both food distribution and the phytochemical properties of foods influenced female social relationships in Hanuman langurs at Ramnagar, with evidence showing that these langurs concentrated their feeding on three resources that were scarce and clumped (Koenig et al. 1998). Two of these three resources were significantly higher in extractable protein and soluble sugar than other food plants. However, data on contest competition, food related or otherwise, were not collected. The linear dominance hierarchy alone was used as evidence of a link between feeding competition and food availability in these langurs (Koenig et al. 1998). Such a conclusion might emerge if the dominance hierarchies of adult female vervets on Segera in different habitats were compared with the foods available in these habitats. Vervet females would not be expected to exhibit a stable, linear dominance hierarchy based on the time they spent feeding/foraging within the whistling-thorn woodland, which is characterized by small, abundant, and scattered feeding sites. However, vervet females would be expected to contest foods at a higher rate in the riverine habitat, which is characterized by larger, abundant, and more widely spaced feeding sites. Besides competing for food, however, adult female vervets in the current study also competed for access to infants, space, grooming partners, and proximity to other individuals, even though the proportion of nonfeeding contests that occurred was significantly less (26%) than the proportion of contests that occurred during feeding (74%).

Although the dominance hierarchy of adult female vervets on Segera was statistically linear based on contests that occurred within the riverine habitat and on food-related agonism, when only food-related agonism in the riverine habitat was considered, the hierarchy was not statistically linear. Almost half of all agonistic interactions that occurred in the riverine habitat were unrelated to feeding competition, and, in order to produce a significantly linear hierarchy based on feeding contests, agonism over foods within both habitats had to be considered. Thus, whereas vervets exhibited a stable, linear hierarchy in a habitat where feeding competition was expected, feeding competition alone did not explain this finding. Other factors influenced the linearity of the adult female vervet dominance hierarchy. These findings run counter to expectations of theories of feeding competition among female primates. Another explanation that has been put forth to explain the existence of stable dominance hierarchies among primates is that such a construct effectively reduces the amount of aggression. Expectations of this theory include higher rates of aggression as well as the escalation of aggression among species that lack well-defined dominance hierarchies (Bernstein 1981).

I have explored a number of alternate explanations for the maintenance of a stable, linear dominance hierarchy among Segera vervet females and have questioned the significance of feeding competition to the monkeys at this site. If the dominance hierarchy of adult female vervets was strongly linked to feeding competition, then, when feeding competition is relaxed, the dominance hierarchy would weaken. For example, female baboons and macaques exhibit stable and linear dominance hierarchies at some sites but not at others (Barton, Byrne, & Whiten 1996; Koenig et al. 1998). Superficially, the differences in the linearity of the hierarchy exhibited in the whistling-thorn woodland versus the fever tree riverine habitats used by vervets on Segera supported this expectation. However, data on feeding competition among female vervets do not provide the link between the ecological variable (i.e., food availability) and female dominance relationships as predicted by the models of female social behavior. Although the argument might be made that the function of the dominance hierarchy is to provide dominant individuals with preferred access to resources during times of need, such as drought or other crises, and should not fluctuate with short-term environmental changes, such data are rare (c.f., Cheney & Seyfarth 1990; Cheney et al. 2004). I return to a discussion of this issue in the final chapter.

# 6

# Comparing Vervet and Patas Monkeys in Whistling-Thorn Woodland

In this chapter, I examine contest competition and dominance among female patas and vervet monkeys as they use the same whistling-thorn woodland habitat. As described by the ecological and socioecological models, dominance relationships among adult females reflect the patterns of contest competition exhibited and will therefore be influenced by patterns of food availability. This expectation is based on the assumption that food-related contest competition will figure prominently in the patterns of agonism exhibited by adult females (Isbell 1991; van Schaik 1989; Wrangham 1980). Vervets and patas monkeys are expected to differ in contest rates and dominance styles based on their characterization in past studies, including published accounts of the groups studied here, but what variables are responsible for producing these differences? Additionally, in order to better understand the link between primate social organization and the foods available to females, we must first understand what specific properties of foods prompt female feeding competition and when. Here, I identify the characteristics of food items that promote contest competition among sympatric vervets and patas monkeys.

## QUESTIONS AND PREDICTIONS

Vervets are classified along with many macaques and most baboons as primate species in which dominance relationships among adult females are nepotistic, despotic, stable, and linear (Isbell & Pruetz 1998). Among patas monkeys, linear dominance hierarchies have been reported only in captive and provisioned animals (Hall 1967; Kaplan & Zucker 1980; Loy & Harnois 1988; Nakagawa 1992). However, the role of dominance in this species has been debated (Goldman & Loy 1997; Isbell & Pruetz 1998; Loy & Harnois 1988; Loy et al. 1993). The lack of submissive signals in patas monkeys has been used as evidence to support a lack of formal dominance in patas monkeys (Loy et al. 1993), although this same argument could be made for vervets (Gartlan & Brain 1968). In a review of dominance relationships among captive patas monkeys during an 8.5-year period, Goldman and Loy (1997) concluded that dominance does not play a significant role in patas monkey intragroup relations. Over the study period, the dominance hierarchies of adult females were linear for short periods (three 1.5-year periods and one 1-year period of stability; Goldman & Loy 1997). Thus, the issue of dominance in patas monkeys is unclear.

I tested the hypothesis that agonistic interactions, especially among adult female monkeys, were influenced by the availability of foods, particularly the primates' main food source, whistling-thorn *Acacia*. I predicted that both species would exhibit similarities in frequency of agonistic interactions when using this food source. The abundance, randomly scattered distribution, and quick food site depletion times of whistling-thorn *Acacia* led me to expect low frequencies of agonistic interactions between adult females in general. I also examined the context in which agonism occurs and its influence on dominance relationships. The severity of agonism may be affected by the context in which it occurs. Specifically, the severity of agonism should, in part, reflect to what degree contested resources are limited. As I noted in the previous chapter, adult females are expected to engage in more intense agonism during feeding than in nonfeeding contests, especially if the food being contested was one high in quality and low in availability. The severity of agonism may also be affected by factors such as social stability. For example, agonism among provisioned adult female patas monkeys was less severe than among immature patas monkeys at Guayacan, Puerto Rico (Zucker 1994). The decrease in severity of agonism with age was suggested to serve a socializing function (Zucker 1994). I expected that, for vervet and patas monkeys, dominance relationships that are less stable will be characterized by agonism that is more severe. In other words, once dominance relationships are established, agonism is expected to be less severe than that characteristic of ambiguous relationships (Bernstein & Ehardt 1985). Loy and co-workers (1993) noted that captive patas monkeys exhibited agonistic rates higher than vervets and baboons (species with "formal dominance") in wide-ranging, large enclosure, and small enclosure situations. If vervets on Segera are "typical" in that they exhibit linear and stable dominance hierarchies (Seyfarth 1980; Struhsaker 1967a,b; Whitten 1983), the intensity of agonistic interactions among adult females is expected to be less severe than that in patas monkeys, a species characterized by a large number of ambiguous dominance relationships.

# AGONISTIC AND DOMINANCE BEHAVIOR

I recorded the occurrences of agonistic behavior for all study group members during focal animal follows (approximately 0730–1600 hours) and *ad libitum* for 4 days per month from July 1993 to December 1994 and, then, opportunistically, from January to August 1995. Data on (1) identity of monkeys involved in the interaction (age, sex, and individual identity), (2) date, (3) intensity of interaction, (4) habitat where interaction occurred, (5) time of day, and (6) reason for or context of interaction (e.g., over food, grooming partner, mating partner, space) were recorded. Dyadic agonistic interactions were then summarized in a dominance matrix. Dominance matrices were used to determine the relative rank of individuals as well as whether the hierarchy was linear and/or stable by calculating the percentages of reversals against the hierarchy and the number of circular relationships.

The overall number of agonistic contests between adult females for both patas monkeys and vervets was low. Agonistic interactions between adult female patas monkeys ($N = 204$) occurred at a rate of 0.36 per hour, whereas adult female vervets interacted agonistically ($N = 77$) at a rate of 0.13 contests per hour when they used the whistling-thorn habitat. When the data were weighted to adjust the number of female vervets to the number of females in the patas monkey group, vervet females were estimated to engage in 0.29 interactions per hour in the whistling-thorn habitat. The rate of agonism exhibited by adult female vervets and patas monkeys after taking into account the number of potential female partners available per group was not significantly different ($\chi^2 = 1.689$, df $= 1, p < 0.50$).

The context in which agonism occurred was similar for vervets and patas monkeys (Figure 6–1). Although both vervet and patas monkey females rarely contested foods ($N = 126$ and 101 contests, respectively), feeding was the most common activity associated with agonistic interactions relative to other contexts. Monkeys also contested one another for grooming partners ($N = 36$ patas monkeys, $N = 8$ vervets), during group resting periods ($N = 7$ patas monkeys, $N = 5$ vervets), and over space ($N = 27$ patas monkeys, $N = 5$ vervets), infants ($N = 2$ patas monkeys, $N = 8$ vervets), and proximity to certain individuals ($N = 4$ patas monkeys, $N = 1$ vervets). Patas monkey females also engaged in agonistic interactions with one another for access to males ($N = 7$) and shade ($N = 5$).

The types of agonism exhibited were similar in some respects for female vervets and patas monkeys. Approach-avoid was the most common agonistic interaction among female patas monkeys and among female vervets (Figure 6–2). Neither species engaged in frequent physical aggression ($N = 3$ or 0.7% for vervets, $N = 8$ or 2% for patas monkeys). All forms of escalated agonism (chase-flee interactions and physical fights) accounted for approximately 23% of agonistic interactions between female patas

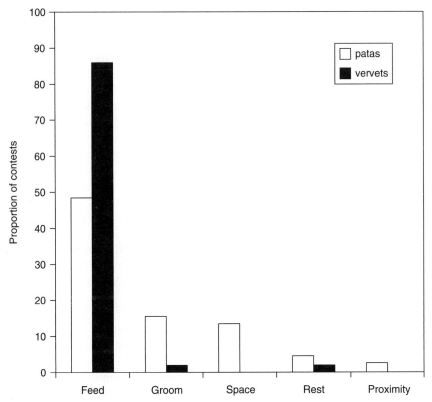

**Figure 6–1** Contexts of agonism for females in whistling-thorn habitat.

monkeys but only 16% of agonistic interactions between female vervets. Of these, only five in the case of patas monkeys and three in the case of vervets occurred in a feeding context. The least intense forms of agonism, such as mounting, cringing, and flinching away, which are classified into one category in Figure 6–3, were rarely observed.

Species differences existed in the proportions of time spent engaged in different types of agonism in certain contexts. During feeding, patas monkeys engaged in a greater number of supplants ($N = 26$) than vervets ($N = 16$, $\chi^2 = 23.81$, df $= 1$, $p < 0.001$). Patas monkey females spent a significantly greater proportion of contests involved in escalated (chase-flee and fight) versus other interactions outside of a feeding context ($\chi^2 = 14.6$, df $= 1$, $p < 0.01$) and in more approach-avoid interactions during feeding contexts ($\chi^2 = 5.1$, df $= 1$, $p < 0.05$) (Figures 6–3 and 6–4). Female vervets engaged in more approach-leave interactions ($\chi^2 = 9.32$, df $= 1$, $p < 0.01$) outside of feeding contexts but in significantly greater approach-avoid interactions while feeding ($\chi^2 = 6.08$, df $= 1$, $p < 0.02$).

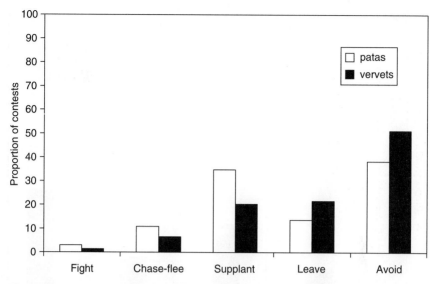

**Figure 6–2**  Intensity of contests among females in whistling-thorn habitat.

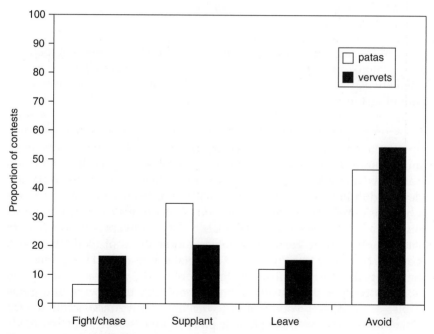

**Figure 6–3**  Intensity of agonism in feeding contexts.

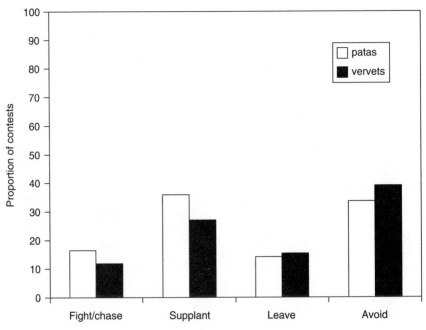

**Figure 6–4** Intensity of contests in nonfeeding contexts.

Given the relatively low occurrence of food-related contest competition among monkeys in my study and the high proportion of all agonism that was between adult females, I decided to consider all age-sex classes of individuals in my analyses in order to increase the sample size of some foods contested and thus the validity of interpretations. This is based on the assumption that although food may not be the most limiting resource, reproductively, for all individuals in a group (i.e., males are reproductively limited moreso by the number of mates available to them), the properties that provoke contests over food are similar across all age-sex classes. Eleven adult females, thirteen adult males, and twenty immature vervets were considered in the analyses of foods that vervets contested, whereas fourteen adult females, three adult males, and over twenty-eight immature patas monkeys were considered subjects in analyses of foods that patas monkeys contested (see Pruetz 1999 for details on male and immature subjects). Vervets competed for whistling-thorn seeds, whereas female patas monkeys competed for swollen thorns, new growth, flowers, and gums (Tables 6–1 and 6–2). Food items of whistling-thorn *Acacia* accounted for 50% of all contests involving female patas monkeys over identified foods (see Table 6–2). This proportion was significantly higher than the proportion of contests over whistling-thorn foods in which vervet females were

**Table 6–1**  Foods Contested by Vervets

| Vervets | Frequency of Contests |
|---|---|
| Mushroom species | 16 |
| *Scutia myrtina* **fruit** | 11 (1) |
| *A. drepanolobium* **unknown food** | 7 |
| Grass blades (unknown species) | 7 |
| Roots (unknown species) | 7 (5) |
| Stalk (unknown species) | 6 (1) |
| *A. drepanolobium* **new growth** | 5 |
| *A. xanthophloea* **gum** | 5 |
| *Brachiara brizantha* **grass** | 4 |
| *Ximenia americana* fruit | 4 |
| Vine fruit (unknown species) | 4 |
| *A. drepanolobium* **seeds** | 3 |
| *A. drepanolobium* **thorns** | 3 |
| *A. drepanolobium* gum | 3 |
| *Lyceum europium* leaves | 3 (2) |
| Termite mound item (mushrooms?) | 3 |
| Termites | 2 |
| *A. seyal* | 2 |
| *A. gerrardii* seeds | 1 |
| *Euclea divinorum* fruit | 1 |
| *A. xanthophloea* **seed pods** | 1 |
| *A. seyal* **immature leaves** | 1 |
| Water | 1 |
| *Grewia similis* fruits | 1 |
| *Ipomoea* sp. flowers | 1 |
| *Asparagus aethiopicus* vegetation | (1) |
| *Commelina* sp. (herb) | (1) |
| *Pennisetum* sp. (grass) | (1) |
| *Sarcostema viminale* (stems) | (1) |
| Shrub (unknown species) | (1) |
| **Total contests over identified foods** | **106** |

Numbers in parentheses are the frequency of total interactions between females. Items in bold indicate the top foods in the diet.

involved ($\chi^2 = 16.0$, df $= 1$, $p < 0.001$). Contests between female patas monkeys over gums accounted for the majority of all contests over whistling-thorn foods (53%). Excluding gums from analyses of contests revealed that other whistling-thorn foods accounted for only 32% of all contests between patas monkeys over identified food items, a proportion similar to that for female vervets ($\chi^2 = 0.358$, df $= 1$, $p > 0.80$).

**Table 6–2** Foods Contested by Patas Monkeys

| *Patas Monkeys* | N = 78 |
|---|---|
| *Acacia drepanolobium* **gum** | 12 (3) |
| Mushroom species | 12 (4) |
| Water | 9 (4) |
| *A. drepanolobium* **unknown food** | 6 (2) |
| *Commelina* **spp. fruit** | 5 |
| *Acacia seyal* **food item** | 4 |
| *A. drepanolobium* **swollen thorn** | 3 |
| Francolin eggs (*Francolinus* sp) | 3 |
| Ants (large, terrestrial species) | 3 (1) |
| Stalks (unknown species) | 3 (1) |
| Agama lizard (*Agama agama*) | 3 (1) |
| *Sarcostema viminale* vegetation | 2 (1) |
| On termite mound (mushrooms?) | 2 |
| *Cucumis aculeatus* vegetation | 2 |
| *A. seyal* **new growth** | 1 |
| *A. drepanolobium* new growth | 1 |
| Root (unknown species) | 1 (1) |
| *Rhynchosia malacophylla* legumes | 1 |
| "J7S" plant herbaceous vegetation | 1 |
| **Moth (Lepidopteron sp.)** | (1) |
| *A. drepanolobium* flowers | (1) |
| *Chenopodium* sp. vegetation | (1) |
| *Eriosema* sp. vegetation | (1) |
| **Total contests over identified foods** | **78** |

Numbers in parentheses are the frequency of total interactions between females. Items in bold indicate the top foods in the diet.

## THE LINK BETWEEN FOOD AVAILABILITY AND COMPETITIVE REGIME

Whereas the sum of agonistic interactions is often used to decipher dominance relationships, these interactions vary in context, not all of which center obviously on food. Agonistic interactions over additional resources, such as mates, space, and grooming partners, introduce greater complexity to female social relationships than simple ecological or socioecological models acknowledge. I examined the specific properties of foods that vervets contested in order to see whether the models of female social behavior would have predicted such contests according to these variables (e.g., abundance and distribution).

I used analyses of variance and multivariate regression to examine the effects of different aspects of food availability on the contest behavior of vervets and patas monkeys (Table 6–3). These included the proportion of the diet that each food item contributed to overall content and the abundance of these foods on two scales: (1) abundance per hectare and (2) abundance per food patch, where a patch was considered the crown in the case of trees and shrubs. Other variables considered included the foods' distribution according to random (patchy, clumped, or evenly dispersed), processing costs, and protein content. The foods included in analyses were the top food items of vervets and patas monkeys (such as *Acacia* gum and swollen thorns; see Isbell et al. 1998a), in addition to several others that monkeys were observed to contest and for which ecological data were available (Isbell & Pruetz 1998). As I noted previously, in order to examine food-related contests specifically, I include data on agonistic interactions over food among all age-sex classes of monkeys in order to increase the sample size for individual foods contested and the statistical validity of my interpretations.

In the analyses of the abundance of foods, data on the number of most of the food items per hectare were available; however, foods that were apparently evenly distributed, such as herbaceous-level foods like fruits of *Commelina*, were not included in the analyses of patch abundance because clear patches of these foods were not discernable by human observers; that is, they constituted "patches" large enough for all group members to feed in simultaneously. In the analyses of food distribution, mushrooms, gums of *Acacia seyal*, flowers of *Acacia xanthophloea*, and leaves of *Lyceum europaeum* were not included in all analyses because data on these food items were not available. Protein content was not available for fruits of *Commelina, Scutia myrtina*, and the insects within swollen thorns.

Analyses of variance tests showed that the foods contested most frequently by vervets and patas monkeys were, for the most part, common in the diet, such as gum of *Acacia* and fruit of *S. myrtina* (Table 6–3). The proportion of the diet that a food item accounted for significantly influenced the frequency at which it was contested. This implies that monkeys were competing over foods in proportion to their inclusion in the diet. Highly contested foods were also clumped in space at all scales, that is, regarding

**Table 6–3** Variables Significantly Increasing Contest Competition Based on ANOVA Tests

| Variable Considered | F-Value | df | p-Value |
| --- | --- | --- | --- |
| Higher proportion in diet | 6.2 | 10 | 0.004 |
| Greater number per hectare | 975.5 | 8 | 0.000 |
| Greater number per patch | 263.7 | 6 | 0.000 |
| Proportion of diet + number per patch | 6.3 | 2 | 0.02 |
| Proportion of diet + number per hectare | 4.6 | 2 | 0.04 |

abundance per food patch and per hectare. Neither the spatial distribution of foods as measured by the PCQ method nor the amount of protein contained by foods significantly affected the frequency with which monkeys contested them.

When several variables were considered together, the proportion of the diet that a food item accounted for as well as the abundance of the food item on the two different scales significantly affected the frequency with which monkeys contested it (see Table 6–3). A multivariate ANOVA showed that patch abundance and proportion of diet best indicated when primates would contest foods, although when proportion of diet was combined with abundance per hectare, there was also a significant effect. In each of these scenarios, proportion of diet of the food item contributed significantly to the equation, whereas the other variables did not. When the proportion of the diet a food accounted for was combined with other variables, such as distribution or protein content, no significant relationships were found.

Ideally, an array of variables that influence feeding competition should be considered in models linking ecology to social behavior. These include the spatial and temporal distribution and abundance of foods, as well as processing costs associated with foods (e.g., ant protection in this study), nutritional benefits, and the relative importance of alternate or additional foods. For example, in this study, the availability of each whistling-thorn food type was considered in relation to other foods of the same species and was then scored as to their contestability. For the different food items, each of the variables was weighted equally in a heuristic model (see Tables 3–8 and 3–9) designed to assess which aspects of food availability were most influential in promoting contest behavior. For both vervets and patas monkeys, I would expect that flowers and gums would be contested more often than other whistling-thorn foods, such as seeds, new growth, and swollen thorns.

Patas monkeys and vervets contested foods that were widely distributed in space, abundant at food sites, and had relatively long food site depletion times (e.g., water, mushrooms growing on termite mounds, and gum of *Acacia*). This does not fit the expectations of the ecological and socioecological models. Gums are not exceptionally high in protein compared to other foods such as seeds, leaves, and flowers. They are high in fiber and carbohydrates and low in tannin content. In terms of processing costs, gum of *Acacia* is not defended by obligate ant species, whereas mushrooms are relatively easily processed compared to most *Acacia* foods (i.e., not protected by thorns or ants). Vervets also contested fruits of *Scutia myrtina*, which were found in widely spaced, abundant food sites (i.e., traditional definition of patchy) and not protected by ants.

An examination of the foods contested by patas monkeys suggests that, in the case of gums of *Acacia*, the predictions that maintain that clumped foods elicit higher rates of contest behavior were accurate. However, the difference was not significant due to the low frequency of contests over foods in general. According to the models, patas monkeys were also expected to contest flowers. However, this was observed only once. Supplants accounted for three contests over gums and a single contest over flowers. All other contests over whistling-thorn foods involved less intense agonism. Gum was more quickly processed by patas monkeys (per bite) and was of higher nutritional quality than flowers. Additionally, whereas both flowers and gum were not protected by obligate ant species, gum is often found on the trunk of whistling-thorn trees and therefore less protected by thorns. Thus, even though flowers were more widely distributed in space and less abundant than gum, they were rarely contested. Additionally, they did not account for a great proportion of patas monkeys' diet during this study, as did gum (Isbell 1998). Other foods that patas monkeys contested included swollen thorns and new growth, which were ranked high nutritionally. However, swollen thorns were ranked lowest in relative abundance and distribution and regarding processing costs, and new growth was ranked lowest regarding its distribution.

## DOMINANCE IN ADULT FEMALES

The dominance hierarchies of vervets and patas monkeys were compared as each species used whistling-thorn woodland. As I noted in the previous chapter, the dominance hierarchy for female vervets was statistically linear and characterized by only one reversal and two unknown relationships. Higher-ranking female vervets directed significantly more agonism than did lower-ranking females but did not receive less agonism than did lower-ranking females. In whistling-thorn woodland, vervets maintained a linear pattern, but the hierarchy was not statistically linear, probably due to the large number of "unknown" relationships in the matrix.

The dominance hierarchy of female patas monkeys was not statistically linear ($d = 90.75$, df = 21.84, $\chi^2 = 22.44$, $p > 0.30$, $K = 0.202$; see Table 6–4). Because the sample size is greater for female patas monkeys, a Chi-square test could be used to determine significance (after Appleby 1983 with groups of $N > 10$), although caution must be exercised because Chi-square tests assume independence of samples (Kramer & Schmidhammer 1992). The dominance hierarchy exhibited by female patas monkeys was characterized by a number of reversals (5%) and many unknown relationships ($N = 40$ blank cells), although females could be ranked into a hierarchy.

Two different analyses were used to explore the effects of rank on adult female patas monkey behavior. First, an overall ranking of high or low was assigned to individual adult females in order to assess whether there were rank-related differences in the amounts of agonism directed and received. The top seven adult females were classified as high ranking, whereas the bottom seven were classified as low ranking. When females

**Table 6–4** Dominance Matrix: Adult Female Patas Monkeys (November 1993–August 1995)

| | geo* | sco* | gya | pen | war | mnt | vnc | taz* | mic | cez | rem | pic | ren | dal |
|---|---|---|---|---|---|---|---|---|---|---|---|---|---|---|
| geo* | | | | | | | 1 | | | | | 2 | | 2 |
| sco* | | | 2 | | | | | 2 | | | | | | |
| gya | | | | 1 | 2 | | 2 | | | | 1 | 2 | | |
| pen | 1 | | | | 1 | | 1 | | 1 | 2 | | 1 | | 2 |
| war | | | | | | 1 | 2 | 2 | 1 | | | 1 | | 1 |
| mnt | 1 | | | | | | 1 | | | 5 | | 2 | | 4 |
| vnc | | | | | | | | 2 | 1 | | 1 | 2 | | 3 |
| taz* | | | | | | | | | 1 | 1 | | 4 | | 4 |
| mic | | | | | | | | | | | | 1 | | |
| cez | | | | | | | | | | | 3 | | 1 | |
| rem | | | | | | | | | | | | | | |
| pic | | | | | | | | | | | | | 1 | 3 |
| ren | | | | | | | | | | | | | | |
| dal | | | | | | | | | 1 | | | | | |

*Included in dyads structurally unable to interact and thus excluded from analyses of linearity.

were categorized this way, significant differences in the amount of agonism that they directed ($Z = -2.941$, $p = 0.002$) and received ($Z = -2.805$, $p = 0.002$) were evident. High-ranking females received significantly less and directed significantly more agonism than low-ranking females. Next, adult females were assigned a rank within the hierarchy from one to fourteen, in descending order from top to bottom of the dominance matrix. A Wilcoxon matched-pairs signed ranks tests showed individual adult female rank did not affect the amount of directed ($Z = 0.094$, $p = 0.463$) or received ($Z = 0.409$, $p = 0.342$) agonism when females were assigned specific, rather than general, ranks.

Food-related agonism was examined in detail because these interactions were expected to accurately reflect the influence of food availability on adult female patas monkey behavior. Results of analyses of food-related agonism and resultant dominance hierarchies among female vervets were presented in the previous chapter. Agonistic interactions that occurred among female patas monkeys in the context of feeding and the resultant dominance hierarchy are given in Table 6–5. The order of females in the hierarchy has been changed in order to reduce the number of reversals. "TAZ" and "CEZ" have moved above "MIC" in the hierarchy. The number of reversals observed during feeding was 2% of all observed interactions, slightly less than the percentage of reversals observed during all contexts. Using Appleby's (1983) modification of Kendall's (1962) method of determining linearity in hierarchies, the female patas monkey dominance

**Table 6–5** Dominance Matrix: Food-Related Contests in Female Patas Monkeys

| | sco | gya | pen | war | mnt | vnc | taz | cez | mic | rem | pic | ren | dal |
|---|---|---|---|---|---|---|---|---|---|---|---|---|---|
| sco | | | 2 | | | | 1 | | | | | | |
| gya | | | 1 | | 1 | | | | | 1 | | | |
| pen | 1 | | | | | | | 1 | 1 | | | | |
| war | | | | | 1 | | | 1 | | | 1 | | 2 |
| mnt | | | | | | | 1 | 2 | 1 | | 1 | | 2 |
| vnc | | | | | | | 2 | 1 | 1 | | 2 | | 3 |
| taz | | | | | | | | 1 | | | | | 1 |
| cez | | | | | | | | 1 | 1 | 1 | | | |
| mic | | | | | | | | | | | | | |
| rem | | | | | | | | | | | | | |
| pic | | | | | | | | | | | | 1 | 1 |
| ren | | | | | | | | | | | | | |
| dal | | | | | | | | | | | | | |

hierarchy based on food-related contests was not significantly linear ($d = 68$, $df = 21.19$, $\chi^2 = 24.75$, $p > 0.29$, $K = 0.253$), and a large number of dyads existed for which dominance relationships were unknown.

Finally, a dominance matrix was constructed that excluded female patas monkeys just maturing and entering the hierarchy in order to remove the potentially confounding variable of social experience (Rothstein 1992). Table 6–6 illustrates the agonistic interactions among adult female patas monkeys who had had an infant prior to the onset of this study in 1993. Adult females less than 3 years old (age at first birth) at the onset of the study were excluded from the analyses. This hierarchy thus represents adult females who had access to one another and were observed for at least 18 months. A considerable number of ambiguous relationships remained, and the hierarchy was not statistically linear. Reversals within the hierarchy only accounted for 3% of interactions, however, so that females could still be ranked linearly.

## FEEDING COMPETITION AND PRIMATES' FOODS

Although patas monkeys and vervets were expected to show similarities in the frequency of agonistic interactions occurring in the same habitat, the two species differed in some respects. Some of these differences can be related to contrasts in patterns of food item distribution and abundance in vervets' and patas monkeys' home ranges. However, overall agonism over food was infrequent. Food items of *Acacia* that were relatively more clumped and less abundant, such as gums, elicited higher frequencies ($N = 3$) of contest competition than food items that were more scattered and abundant, such

**Table 6–6** Dominance Matrix for Adult Female Patas Monkeys >3 Years of Age

|     | gya | pen | war | mnt | vnc | mic | cez | rem | pic | ren | dal |
|-----|-----|-----|-----|-----|-----|-----|-----|-----|-----|-----|-----|
| gya |     | 1   | 3   |     | 2   |     |     | 1   | 2   |     |     |
| pen |     |     | 1   |     | 1   | 1   | 2   |     | 1   |     | 2   |
| war | 1   |     |     | 1   | 2   | 1   |     |     | 1   |     | 2   |
| mnt |     |     |     |     |     | 1   | 5   |     | 4   |     | 2   |
| vnc |     |     |     |     |     | 1   | 1   | 1   | 3   |     | 4   |
| mic |     |     |     |     |     |     |     |     | 1   |     | 1   |
| cez |     |     |     |     |     |     |     | 4   | 1   |     |     |
| rem |     |     |     |     |     |     |     |     |     |     |     |
| pic |     |     |     |     |     |     |     |     |     | 1   | 4   |
| ren |     |     |     |     |     |     |     |     |     |     |     |
| dal |     |     |     |     | 1   |     |     |     |     |     |     |

as flowers ($N = 1$), and gum was relatively more clumped within patas monkeys' home range than within vervets' home range. Additionally, gum was the only whistling-thorn food that seemed to be depleted by patas monkeys at individual feeding sites. When agonistic interactions involving gums were excluded from analyses, the rate of agonism over whistling-thorn food items was similar for patas monkeys and vervets.

Even though patas monkeys and vervets had access to whistling-thorn trees of different heights, because the ants' protection of foods on *Acacia* restricted the monkeys to a few items per tree, the importance of tree height in predicting the availability of certain foods (those protected by ants) was reduced. According to the models, since trees in the patas monkeys' home range were scarcer and more widely spaced than those available to vervets, competition between patas monkeys over whistling-thorn foods should be greater. However, both species showed similar levels of contest competition over most whistling-thorn foods. The fact that patas monkeys range further per day than vervets (Hall 1965; Kavanagh 1978; Nakagawa 1989b) seems to equalize the effects that the differences in distribution of whistling thorn in these two species home ranges have on the monkeys' contest behavior. Patas monkeys' ranging patterns enables them to use a large number of trees per day, even though the trees in their home range were more widely spaced than those in the vervets' home range.

Based on the characteristics of foods discussed here, I compiled a table (Table 6–7) of conditions proposed to elicit contest competition among female primates. Although processing costs and the availability of other foods were also influential factors in determining what foods females contest, general predictions of contest competition were supported with data on the distribution and abundance of foods in combination with data on primates' dietary preferences. In other words, primate feeding behavior

**Table 6–7** Patterns of Food Availability and Consequent Patterns of Female Contest Competition in Vervets and Patas Monkeys

| |
|---|
| **Higher degree of contest competition** |
| Clumped, seasonally abundant (e.g., *Scutia* fruit) |
| Clumped, generally scarce (e.g., mushrooms) |
| Scattered, scarce but preferred (e.g., gums) |
| Widely spaced sites (e.g., *Scutia* fruit) |
| Processing costs expected to be low, although they may be high for preferred foods that are scarce |
| Nutritional benefits may be high–medium but not relatively low |
| **Lower degree of contest competition** |
| Randomly scattered, superabundant |
| Evenly distributed, superabundant, or abundant |
| Clumped, superabundant |
| Processing costs relatively high |
| Nutritional benefits relatively low |

with respect to foods' nutritional content is a key factor determining when primates should contest foods and may override other variables (e.g., in the case of whistling-thorn *Acacia*). Thus, I predict that even foods that are scattered may be contested when they are relatively scarce; processing costs are expected to be low for such foods for the most part, although, in the case of preferred foods, processing costs could be high; nutritional costs are expected to be generally high. I included foods with long food site depletion times that were also widely separated in space as foods likely to elicit strong contest competition. Foods not expected to elicit strong competition include those randomly distributed and superabundant, evenly distributed and abundant or superabundant, or evenly clumped but superabundant. High processing costs and low nutritional costs also contribute to the expected low degree of contest competition characteristic of these types of foods.

The importance of dominance to female vervets and patas monkeys appears to differ in some respects. All female vervets, except the alpha female, received some agonism from other adult females. Although high-ranking female vervets directed significantly more aggression at other adult females, there was no difference in the amount of agonism received by either high- or low-ranking females. In this case, vervet female rank in the dominance hierarchy did not seem to serve to protect against aggression, except in the case of the alpha female.

I demonstrated in the previous chapter that the dominance hierarchy of female vervets in the whistling-thorn habitat was not significantly linear. This was probably due to the large number of unknown relationships. Although the rate of contests for vervets in whistling-thorn woodland was lower than in riverine habitat, no reversals against the hierarchy occurred for adult female vervets here, suggesting that the dominance relationships that existed among adult females were stable. However, even though the hierarchy of female vervets did not break down when they used less usurpable food resources, contest rates indicate the decreasing importance of competition in a habitat characterized by widely scattered foods that occurred in small patches. In the riverine habitat, the main vervet food source, fever tree *Acacia*, was more clumped in distribution and more abundant at food sites (i.e., longer food site depletion times) than whistling-thorn *Acacia*. The foods available within the riverine habitat may so strongly influence the hierarchy of female vervets that, although contest competition decreases in the whistling-thorn woodland, rank reversals do not occur (Isbell & Pruetz 1998; Isbell et al. 1998b). If food is limiting to female vervets in some context, high-ranking females are expected to maintain their dominance rank in all contexts because rank maintenance is expected to be less costly for females than reestablishing rank order during times of limited food availability. For example, costs related to escalated aggression in contests where rank is uncertain would be higher than during contests where rank is established and, thus, aggression rarely escalates. The maintenance of dominance rank may also reflect a strategy that anticipates times of resource scarcity, so that maintaining high rank would be advantageous for the same reasons noted above. Cheney and co-workers (2004) suggest this possibility. Their 10-year study of chacma baboons in Botswana revealed that females' lifetime reproductive success is largely similar, regardless of rank, although specific causes of mortality are rank related. For example, while interbirth interval and infant growth rates were shorter and faster, respectively, for high-ranking female baboons, these same females suffered higher rates of infant mortality due to infanticide. They suggest that benefits of being high ranking for adult female chacma baboons are most likely to be seen under conditions of extreme ecological conditions, such as in times of drought or food scarcity (Cheney et al. 2004).

Dominance relationships among patas monkeys were more variable than among vervets. Adult female patas monkeys exhibited more escalated forms of agonism ($N = 26$) than vervets ($N = 10$), for example. Adult female patas monkey dominance relationships were not as delineated as those reported for provisioned patas monkeys, free-ranging vervets, and other cercopithecines, such as baboons and macaques. In a study of langurs (*Presbytis entellus*) at Jodhpur, Borries (1993) described the dominance hierarchy of adult females as unstable compared to macaques and baboons, based on the proportion of reversals (3.9% of displacements) and changes in rank (2.7% of displacements). Similarly, 5% of agonistic interactions among adult female patas monkeys in my study did not conform to a linear hierarchy.

The combination of traits characteristic of female patas monkey dominance includes a high proportion of escalated aggression, rank effects on directing and receiving agonism, and a linear dominance hierarchy that is unstable and not significantly linear. The fact that escalated contests accounted for over 12% of all agonism witnessed among adult females is consistent for species in which a dominance hierarchy is not stable. Hall (1967) found that aggressive interactions among patas monkeys were rare, and physical fights were never observed during his 6-month study of patas monkeys in Uganda. In the current study, physical aggression among female patas monkeys also was extremely low, instances of wounding in this species were less common compared to vervets (personal observation), and the proportion of contact aggression among female patas monkeys (0.7%) was low. The rank effects observed in patas monkeys, with high-ranking females receiving significantly less aggression (mean 4.3 aggressive acts) than low-ranking females (mean 12.7 aggressive acts), do not correspond with the suggestion that dominance is unimportant in this species.

It has been suggested that, for many primate species, agonism is most common during feeding, and rank is thought to play an important role in determining females' access to foods. Among vervets in Samburu, Kenya, adult females most often disputed access to food items (Whitten 1984). Hall (1967) noted that captive adult female patas monkeys positioned themselves near the center of the feeding space (occupied by the adult male of the group) based on their relative rank within the group. Sterck and Steenbeek (1997) found that dominance relationships among adult female Thomas langurs were more differentiated inside food patches rather than outside food patches. Feeding patches used by patas monkeys are dispersed and small, with quick depletion times (Isbell & Pruetz 1998; Isbell et al. 1998b and this study), whereas those used by langurs are larger relative to adult female body size (Sterck & Steenbeek 1997), so that one might conclude that the foods available to patas monkeys and langurs directly influence the types of dominance relationships they exhibit. However, considering that the number of adult females langurs in Sterck and Steenbeek's (1997) study averaged three females per group ($N = 3$ groups), these authors' conclusions cannot be statistically assessed. According to Appleby (1983), a dominance hierarchy in a social group of less than six individuals (from whom the dominance matrix is calculated) cannot exhibit a statistically significant level of linearity. Groups with a sample size of three are reported to have a 75% probability of exhibiting linearity or near linearity due to chance alone (Appleby 1983). This example emphasizes the problems inherent in asserting that, for example, patas monkeys' lack of a linear dominance hierarchy is caused by the patterns of food resources available to them, whereas Thomas langurs exhibit a statistically linear hierarchy because of the nature of their foods. The frequency of contest behavior of patas monkeys indicates that food-related competition (less than half of all agonistic contests among adult females) played only a minor role in the type of dominance hierarchy they exhibit.

I suggest that the long-term instability of the female hierarchy among patas monkeys (Isbell & Pruetz 1998) may be related to the life-history patterns of this species (Rowell & Richards 1979) more so than to the availability of their food resources. Patas monkey females mature at a relatively early age. Primiparous females usually conceive at 2.5 years and give birth to their first offspring at age 3 (Chism & Rowell 1988). The interbirth interval of patas monkeys in Laikipia is approximately 1 year (Chism & Rowell 1988), being relatively short in comparison with most cercopithecines (Butynski 1988). Combined with the high level of adult female mortality found for this study group (Isbell & Enstam 2002), the dynamic social group membership of maturing females may contribute to the lack of stability. Adult female patas monkeys exhibit a hierarchy with a linear pattern, so that females could be ranked, but one that is not statistically linear or stable over the long term. Because, on Segera, some nulliparous females matured and gave birth every year (see also Carlson & Isbell 2001), it is not surprising that the hierarchy would fluctuate over such short periods of time. Rank-related differences in the amount of agonism directed and received, however, indicate that short-term dominance could affect females' reproductive success. Enstam and co-workers (2002) noted that the most subordinate female patas monkey in the same study group on Segera was the victim of what was interpreted as male-directed infanticide of her 3-month-old infant. The most dominant female at the time was the only other lactating female, and the newly resident male was not observed to target her infant.

## SIGNIFICANCE OF FEEDING CONTEST COMPETITION TO VERVETS AND PATAS MONKEYS

The models presented to explain female social behavior among primates assume that when food is limiting, adult females actively compete for access to productive feeding sites (Isbell 1991; van Schaik 1989; Sterck, Watts, & van Schaik 1997; Wrangham 1980). More precisely, however, females should compete when usurping food resources from others confer an advantage to the usurper, such as reduced travel to and search costs for alternate food sites (Pruetz & Isbell 2000). The individual from whom food is usurped may suffer costs in time and energy only if food is limited and distances between adjacent feeding sites are 100s to 1,000s of meters apart. In species that exhibit stable and linear dominance hierarchies, this variability among individuals in such costs and benefits theoretically leads to differences in reproductive success among females of different rank (Wrangham 1980). The social behavior of adult females (i.e., dominance relations), consequently, may be strongly influenced by feeding competition.

Vervets and patas monkeys exhibited lower rates of agonism than vervets at other sites as well as many cercopithecines and other primates for which contest competition has been measured. I attribute this low rate of agonism to the foods available on Segera. The whistling-thorn woodland is characterized by foods that are randomly scattered and abundant, and food patches are small. Such a low rate of agonism implies that feeding competition is less important on Segera compared to other sites where vervets have been studied (e.g., Amboseli & Samburu, Kenya). If food were not limiting for monkeys on Segera, levels of food-related contest competition should be low (van Schaik 1989). Taking into account the proportion of agonistic interactions that were related to feeding, however, implies that food may be the most limiting resource for monkeys on Segera. The rate of agonism exhibited by female vervets during feeding was significantly higher than that during nonfeeding, and patas monkeys also contested more often while feeding than during other contexts. Furthermore, the dominance hierarchy of female vervets based on food-related agonism was significantly linear. However, a dominance hierarchy based on food-related dyadic agonism was not significantly linear for patas monkeys. Taken together, these results suggest that feeding competition plays a more important role in shaping female vervet dominance relationships than in shaping female patas monkey dominance relationships. The demographic differences between the two species may also mask the importance of feeding competition for patas monkeys if my interpretations are correct regarding the absence of a linear, stable dominance hierarchy in this species. While it is not surprising that patas monkeys do not exhibit linear, stable hierarchies given the high growth and mortality rate of this species on Segera, feeding competition may still play an important role in within-group dynamics. However, the degree to which contests over food relate to female survival or reproductive success remains unclear.

# 7

# Testing Predictions of Models of Female Primate Behavior and Ecology

In my study, I examined several models that have been proposed to explain and predict female primate social behavior (Isbell 1991; van Schaik 1989; Sterck, Watts, & van Schaik 1997; Wrangham 1980). Here, I evaluate the accuracy of these models in predicting the female social relationships characteristic of vervets and patas monkeys on Segera, based on the data collected in my study. In particular, I assess the utility of using broad categories of food availability to predict female social relationships in these species. I review the ways in which specific characteristics of foods and primates' feeding behavior influenced contest competition among vervets and patas monkeys and, subsequently, how patterns of feeding competition influenced adult female dominance relationships in these species. Although a number of models use food availability to predict the competitive regimes of female primates living in groups and, subsequently, social organization, data used to support the underlying theory are complex, and patterns of food distribution and abundance are inadequately measured for most primates. Additionally, female primates contest one another over resources as varied as mates, space, and grooming partners (e.g., Schino 2001).

## FEMALE CONTEST COMPETITION AND DOMINANCE IN VERVETS AND PATAS MONKEYS ON SEGERA

Central to the models examined here are assumptions regarding contests over foods. If food-related agonism can be separated from interactions related to "social dominance" (see Isabirye-Basuta 1988) to give a reliable indicator of the importance (or the unimportance) of food availability to primates, supplants over foods should be the most reliable indicator of the value of the contested food resource to the usurper (but see Johnson 1989). When a resource is usurped from one individual by another, the usurper gains direct benefits. The loser may incur a cost if food is limited or widely scattered. The loser may incur no cost if food is not limited or distributed in nearby patches. The supplant rates for adult female vervets in Samburu, Kenya (Whitten 1982, 1983), were similar to those of patas monkeys on Segera but higher than those of Segera vervets. Adult female langurs at Abu, Rajasthan (Hrdy 1977), supplanted one another at higher rates than both vervets and patas monkeys on Segera, as did adult female olive baboons in Laikipia, Kenya (Barton, Byrne, & Whiten 1996) (Table 7–1). Thomas langur and lion-tailed macaque females at Ketambe, Indonesia, were similar regarding their supplant rates outside of food patches compared to the Segera primates (Sterck & Steenbeek 1997) but had higher rates of supplants within food patches. Chacma baboon females exhibited supplant rates similar to those of vervets on Segera (Barton, Byrne, & Whiten 1996). Low supplant rates could indicate that dominance relationships are stable and that individuals employ an avoidance strategy, or they could indicate that resources are not worth usurping (e.g., abundant, evenly dispersed).

Predictions of contest competition among adult females are often based on the size and/or distribution of food patches relative to social group size (van Schaik 1989; Sterck & Steenbeek 1997; Symington 1988). Feeding sites that do not permit all group members to feed simultaneously are expected to promote contest competition (van Schaik 1989) if group members are unable to feed on alternate foods, to feed at nearby sites, or to feed sequentially at the same site after other individuals have left a patch. For example, based on individual tree dimensions, whistling-thorn food sites are small relative to group size for the monkeys in this study (after Sterck & Steenbeek 1997). On the basis of feeding site size alone, I would expect whistling thorn to be contested. The behavior of obligate *Acacia* ants reduced these patches' contestability, however. Additionally, whistling-thorn trees are abundant within the home range of both vervets and patas monkeys and are closely spaced. Not surprisingly, therefore, little competition occurred overall regarding whistling-thorn feeding sites among monkeys on Segera.

Although the frequency of contests among monkeys on Segera was low in general, the proportion of agonistic contests involving whistling-thorn food items was higher than expected compared to other foods. Whistling-thorn food items accounted for almost 20% of identified foods that were contested by vervets, whereas foods of fever tree *Acacia* accounted for 6% of identified contested foods (see Table 6–1). Fever tree *Acacia* was less abundant for the vervets on Segera, based on crown volume values, compared to whistling-thorn *Acacia*. Fever tree feeding sites were larger and more widely spaced than whistling-thorn food sites, however. The number of contests over whistling-thorn as well as the abundance of whistling-thorn crown

**Table 7–1**  Food-Related Agonism in Some Free-Ranging Primate Species

| Species and Source | Hourly Rate, % Contests | Interaction Type and Individuals Involved* | Linear Hierarchy | Sample Method |
|---|---|---|---|---|
| E. patas (1) | 0.33, 49 | See Table 5–1, AF | No | Ad lib |
| C. aethiops (1) | 0.21, 74 | See Table 5–1, AF | Yes | Ad lib |
| C. aethiops (7) | 10.5% | Approach-retreat interactions; AF | Yes | Focal+ ad lib |
| C. mitis (15) | 82% | Aggressive interactions, approach-retreat, spontaneous submission | No | Focal + ad lib |
| P. ursinus (2) | 0.07, 6 | Threat, submission, supplant, chase, fight; AF | Yes± | Focal + ad lib |
| Cercocebus torquatus (16) | 0.59 | Agonism, including yield, avoid, crouch, stare, stare, and lunge, fight | Yes | Focal + ad lib |
| Macaca fuscata (3) | 0.75, 80 | Attack, threat, submission; AF | | Focal |
| M. fuscata (9) | 0.19, 17 | Supplant; aggression: threat, chase, push | Yes? (implied) | Focal |
| M. maurus (13) | 0.20, 23 | Aggression: threat, slap, grab, bite, chase >2 m, AF | Yes± | Focal AF |
| M. fascicularis (11) | 0.56, 32 | Displacements; AF | No,†: two groups | Focal |
| Presbytis entellus (3) | 48.8% | Displacements; AF | Not stable | |
| P. thomasi (11) | 0.46, 72 | Displacements; AF | No and† | Focal |
| Trachypithecus phayrei (17) | 68 and 40% | Aggression, submission and displacements; AF in two groups | Yes and† | Focal + ad lib |
| Procolobus badius (13) | 0.24 | Agonistic: threat, push, bite, hit, chase, steal food, yield, flee, crouch; AF | ? | Focal + ad lib |
| Colobus polykomos (13) | 0.78 | Agonistic: threat, push, bite, hit, chase, steal food, yield, flee, crouch; AF | Yes | Focal + ad lib |
| Pan troglodytes (15) | 51% | Avoid/submission; AF | Yes | |
| Gorilla gorilla beringei (5) | ~43% | Aggression, approach-retreat; AF in six groups | Yes—three groups | Focal + ad lib |
| Ateles paniscus (12) | 0.21, 57 | Agonism in fruit trees | High or low rank | Focal? |
| Saimiri oerstedi (6) | 0.004 | Resource-based aggression; AF | No± | |
| S. sciureus (6) | 0.286 | Resource-based aggression; AF | Yes± | Ad lib |

*AF, adult females; ± indicates that no statistics were performed to determine linearity of the hierarchy; †sample sizes precluded statistical testing.

Sources: (1), This study; (2), Ron, Henzi, and Motro (1996); (3), Saito (1996); (5), Watts (1994); (6), Mitchell, Boinski, and van Schaik (1991); (7), Seyfarth (1980); (8), cited in Borries (1993); (9), Ihobe (1989); (10), Matsumura (1998); (11), Sterck and Steenbeek (1997); (12), Symington (1988); (13), Korstjens (2001); (14), Cords (2000); (15), Wittig and Boesch (2002); (16), Range and Noe (2002); (17), Koenig and co-workers (2004).

area nevertheless indicates that vervets contested whistling-thorn foods proportionately more than expected. Thus, patch size alone was not sufficient to predict the occurrence of feeding contest competition for vervets. An additional consideration, however, is the potential cost associated with contesting fever tree versus whistling-thorn foods. Fever tree feeding sites require individuals to be arboreal, whereas whistling-thorn foods were often eaten from the ground. Barton (1993) found that supplant rates of olive baboons in Laikipia, Kenya, were lower in trees than on the ground and attributed this finding to the increased time and energy involved in arboreal contests, especially those occurring in *Acacia* trees, due to thorns.

It has been suggested that food sites intermediate in size relative to primate social group size are those that should most often be contested (Sterck & Steenbeek 1997) because they preclude all group members from feeding simultaneously but provide a greater return than smaller feeding sites. Fever trees are examples of food sites of intermediate size relative to social group size for vervets in this study. To estimate the number of monkeys that could feed simultaneously at a feeding site, I used calculations of a "feeding sphere" for male and female vervets mentioned earlier based on a hypothetical "approach radius" of approximately 1.0 m for adult female vervets and 1.2 m for adult male vervets, based on their approximate body lengths. Subsequently, using the formula for the volume of a sphere, the "feeding sphere" size for adult vervets was 4.2 m$^3$ for females and 7.3 m$^3$ for males. On average, nine adult males and eight adult females comprised the vervet study group (Pond group). If all adult group members were to feed simultaneously in a fever tree crown, the crown volume necessary would equal approximately 100 m$^3$. This area is about 70% larger than the average crown volume of individual fever trees in the vervets' home range, so only about two-thirds of all adult social group members could feed simultaneously in such a patch. On the basis of predictions made by the models of female social behavior, I would therefore expect that fever tree feeding sites would be contested more than whistling-thorn food sites. Vervets did contest foods proportionately more often in riverine habitat compared to woodland habitat. However, fever tree foods were rarely contested. Because fever trees were distributed linearly and adjacently along the river's edge, rather than contesting such foods, individuals could have fed in adjacent or nearby trees simultaneously, or they could have fed sequentially (see Phillips 1995a,b).

*Scutia myrtina* shrubs could also be considered intermediate-sized food patches for vervets on Segera. The average crown volume of individual *S. myrtina* shrubs was 31 m$^3$, a size scarcely permitting one-third of adult vervets in the Pond group to feed simultaneously. These shrubs were even more widely spaced than fever tree food patches and a relatively smaller patch than fever tree *Acacia* but, compared to whistling-thorn *Acacia*, a larger and more clumped food patch. Approximately 10% of all vervet contests were over *Scutia*, a proportion exceeding that of fever tree *Acacia* but still much less than that for whistling thorn. Overall, within the vervets' range, approximately 10,400 m$^3$ of *Scutia* crown were available compared to more than 41,750 m$^3$ for fever tree *Acacia*. Thus, *Scutia* was contested proportionately more than expected given its inclusion in the diet and its availability to vervets. This particular food was more seasonally restricted than fever tree *Acacia* as well (Pruetz 1999). However, even combining the number of contests over clumped resources, such as fever tree *Acacia, Brachiaria brizantha*, and *S. myrtina*, does not produce the expected contest behavior that would be predicted for vervets on Segera by models of female primate social behavior. Together, these foods accounted for 20% of contests between vervets over identified foods, a proportion that does not differ in respect to that of contests over the superabundant, randomly distributed, and closely spaced whistling-thorn food source. Theories using data on group size relative to food availability (specifically, patch size) did not accurately predict when feeding competition should occur in this case.

In classifying contestable foods, an important factor that must be considered is the relative availability of foods. For example, whistling-thorn gum would not appear contestable for vervets in comparison with other foods because it is an abundant food source. When this food was considered relative to other patas monkey foods, however, it was contestable for that species, being a clumped and locally abundant food source. Additionally, in a survey of 97 gum patches in twelve whistling-thorn trees, only about half of all patches could be described as soft and with a flexible consistency (Pruetz 1999). The other 50% were hard, dark in color, and crumbly. Patas monkeys appear to selectively choose patches that are fresher, therefore reducing the amount of gum available to them within their home range by half, that is, actual versus potential harvest. Patas monkeys also competed over a food that is seemingly abundant in space but seasonally short lived, that is, fruits of *Commelina*. If this food is considered on a different scale, for example, as a temporally patchy resource, the pattern of behavior conforms to models of female primate social behavior. A problem with my analyses of food characteristics was that not all data were comparable for all foods. For example, mushrooms could not be considered in analyses of abundance. Monkeys virtually depleted mushroom patches when they were discovered. Mushrooms occurred in the vicinity of termite mounds, which were usually 3–4 m in diameter and seemed to be widely spaced and scarce within both species' home ranges. A qualitative assessment of this food source is that it is patchy, with long food site depletion times and would likely support the findings that locally abundant food patches were most often contested. Additionally, other aspects of nutritional quality may significantly affect the rate at which primates contest foods, including carbohydrate, fiber, mineral, and tannin content. Although a recent study by Chapman and co-workers (2002) showed that, of these variables, only protein significantly affected diet selection in red colobus. Nakagawa (2003) also used protein as one of the main quality indicators in a comparative study of patas monkey and vervet diet, specifically assessing the protein to fiber ratio of important foods for these primates.

## HOW DO THE MODELS RATE?

The models that have been put forth to explain female social relationships in primates did not consistently predict when vervet and patas monkeys would contest foods. Foods that were important in the diet of these primates accounted for most of the agonism observed over foods, and some of these foods were clumped in space and time (e.g., fruits of *Scutia*). However, contests occurred even over foods that were abundant and randomly distributed in space (e.g., over whistling-thorn swollen thorns and new growth), suggesting that monkeys take advantage of opportunities to usurp foods from subordinate individuals regardless of patterns of food availability. These contests may fall under the classification of "random" contests that serve to maintain dominance hierarchies within the group (Silk 2003). Whereas such contests may seem insignificant, for vervets in this study contests over foods that were abundant and randomly distributed in space and those that occurred outside of a feeding context were important statistically to the stable, linear dominance hierarchy exhibited by adult females. Considering only contests over foods that fit the predictions of models of female primate social behavior (i.e., clumped and locally abundant) did not produce a significantly linear dominance hierarchy among female vervets in this study (Pruetz 1999; Pruetz & Isbell 1998), although the linear pattern prevailed.

The agonistic behavior of adult female vervets and patas monkeys on Segera was not characteristic of cercopithecine species in general (see Table 7–1). Only females in certain populations of Japanese macaques and Moor macaques, during feeding, exhibited rates of agonism similar to monkeys on Segera (Matsumura 1998; Saito 1996). Female moor macaques are characterized as exhibiting relaxed or egalitarian dominance relations, in contrast to other macaque species, even though the hierarchy exhibited by females was stable and linear (Matsumura 1998). Most primates' rates of agonism exceeded rates exhibited by the monkeys studied here, with the exception of bonobos (Furuichi 1997), blue monkeys (Cords 2000), Costa Rican squirrel monkeys (Mitchell, Boinski, & van Schaik 1991), tamarins (Garber 1997), and brown capuchins (Janson 1985). The low rates of agonism in vervets and patas monkeys on Segera are comparable to those between adult female chimpanzees (Wrangham, Clark, & Isabirye-Basuta 1992), provisioned Hanuman langurs (Hrdy 1977), adult female howling monkeys (Zucker & Clarke 1998), white-faced capuchins (*Cebus capucinus*) (Fedigan 1993), and South American squirrel monkeys (Mitchell, Boinski, & van Schaik 1991). Except for white-faced capuchins, these species are not female resident and not usually characterized by strong within-group competition (van Schaik 1989). Even in species in which females are reported to emigrate from their natal group, dominance rank has been reported to correlate with reproductive success (e.g., chimpanzees: Pusey, Williams, & Goodall 1997 but see Kerr 1998). Such differences related to rank are thought to be indicative of stable, linear dominance hierarchies (van Schaik 1989; Wrangham 1980). Koenig and co-workers (2004) suggest that species characterized by female dispersal and clear dominance relations should be considered in models of female social behavior. Clearly, competition, dominance rank, and reproductive success vary among primate species and populations and may not be accurately predicted according to simple models. Additionally, the data provided here focus on species for which agonistic relationships have been explicitly studied, namely anthropoid primates, and do not take into account patterns exhibited by prosimians. Kappeler (1997) notes that the ecological and socioecological models may be especially weak when they are applied to the Strepsirrhines (Table 7–2).

Isbell and Young (2002) note that current socioecological models assume that agonistic interactions, especially those over food, are a relatively common occurrence in species for which dominance is an important organizing feature of society. However, my examination of the proportion of all contests that are food related in vervets and patas monkeys suggests that many contests could also serve to maintain dominance relations rather than directly providing females immediate access to resources (e.g., Silk 2003) or reflect the importance of other limiting resources. An examination of the available literature on contest competition in female primates reveals that, for over half of the species, food-related agonism accounted for less than the majority of all contests observed, which does not meet assumptions of the earlier models of female primate social behavior (van Schaik 1989; Wrangham 1980; see Table 7–2). A similar pattern is revealed in species for which dominance is thought to be an important organizing feature of society based on the nature of these primates' foods. In only a few species the proportion of contests that were food related, the availability of major food resources, and the nature of female competition within groups conformed to the current models' predictions. *Presbytis entellus* conformed to assumptions of the models, but the cercopithecines in general did not. In the three macaque species, food-related contests accounted for less than one-third of agonism, weakening any further correlation between food availability and female competitive regime. In the case of Thomas' langur, for example, females did not exhibit linear dominance hierarchies (although sample size precludes statistical testing), but food-related contests represented 72% of all supplants and displacements, and the hourly rate of agonism was similar to that summarized here for cercopithecids. In most cases, females could be ranked into a pattern consistent with a linear hierarchy even though food-related contests counted for <50% of all

**Table 7–2** Rates of Agonistic Interactions in Wild and Semi-Free-Ranging Female Primates

| Species and Source | Hourly Rate | Type of Interaction | Subjects ± | Reversals | Sample Method |
|---|---|---|---|---|---|
| E. patas (1) | (1) 0.36 | (1) Acacia woodland | AF | (2) 5% | Focal + ad lib |
| | (2) 0.12 | (2) Supplants only | | | |
| C. aethiops (1) | (1) 0.13 | (1) Acacia woodland | AF | (2) 1.3% | Focal + ad lib |
| | (2) 0.04 | (2) Supplants only | | | |
| C. aethiops (19) | 0.79 | Aggression, deference, submission | AF | 0.6% | Ad lib |
| C. aethiops (4, 9) | 0.17 | Supplants over food and space | AF | 1.1% | Focal |
| C. mitis (27) | 0.085 | Aggression: attack, chase, threaten | AF + SAF | | Focal + ad lib |
| Papio anubis (13) | (1) 1.80 | (1) Feeding supplants | AF | | Focal |
| | (2) 0.69 | (2) All supplants | | | |
| P. cynocephalus (17) | 1.80 | All aggression and submission | AF at midcycle | 0.7% | Focal |
| P. ursinus (20) | (1) 1.21 | (1) Approach-retreat | AF | 2.1% | Focal + sequence |
| | (2) 0.05 | (2) "Active aggression" | | | |
| P. ursinus (13) | (1) 0.05 | (1) Feeding supplants | AF | | Focal |
| | (2) 0.03 | (2) All supplants | | | |
| P. ursinus (15) | (1) 1.13 | (1) Before group fission | AF | | Focal + ad lib |
| | (2) 2.28 | (2) After group fission | | | |
| Macaca fuscata (11) | 0.19 | Attack, threat, and submission | AF | | Focal |
| M. fuscata (26) | 1.40 | All agonism | AF | 3.6% | Focal |
| M. maurus (24) | 0.20 | Aggression: threat, slap, grab, bite, chase >2 m | AF feeding | 0% | Focal |
| M. fascicularis (25) | (1) 0.56 | Displacements (1) in food patch (2) outside | AF | 3%, 0% | Focal |
| | (2) 0.19 | | | | |
| M. mulatta provisioned (10) | ~2.36 | All agonistic interactions | AF | 3.4% | |
| Presbytis thomasi (25) | (1) 0.46 | Displacements (1) in food patch, (2) outside | AF | 0%, 5% 15% | Focal |
| | (2) 0.08 | | | | |

| Species and Source | Hourly Rate | Type of Interaction | Subjects ± | Reversals | Sample Method |
|---|---|---|---|---|---|
| *P. entellus* provisioned (8) | (1) 0.28 | Displacements in (1) natural conditions, (2) artificial conditions | AF | 2.6%? | |
| | (2) 0.63 | | | | |
| *Trachypithecus phayrei* (28) | <0.25 | Aggression, submission, and displacements | AF in two groups | 12 and 0% | Focal + *ad lib* |
| *Gorilla gorilla beringei* (18) | (1) 0.93 | (1) All aggression | AF | | Focal + *ad lib* |
| | (2) 0.46 | (2) Scream matches, fights, displays | | | |
| *Pan troglodytes* (14) | 0.15 | Charge, supplant, submissive approach, and response | AF | | Focal |
| *P. paniscus* (16) | 0.00085 | Aggression or submission | AF | | *Ad lib* |
| *Cebus capucinus* (22) | 0.36 | Supplant, "clear" aggression + submission | AF | | Focal + *ad lib* |
| *C. apella* (5) | 0.004* | Feeding displacement; fight, chase, lunge | AF | 2.7% | |
| *Saimiri oerstedi* (23) | 0.004 | Resource-based aggression | AF | | |
| *S. sciureus* (23) | 0.29 | Resource-based aggression | AF | | *Ad lib* |
| *Alouatta palliata* (6) | 0.39 | Displacement, chase, physical aggression | AF | | Focal + *ad lib* |

*Group hour = duration of time period × proportion of group members feeding; ± see text for definitions of agonistic interactions: involves submission of some form; ±AF = adult female × adult female unless otherwise indicated, SAF = subadult female.

Sources: (1), This study; (2), Kaplan and Zucker (1980); (3), Zucker (1994); (4), Whitten (1982); (5), Janson (1985); (6), Zucker and Clarke (1998); (8), Hrdy (1977); (9), Whitten (1993); (10), Sade (1967); (11), Saito (1996); (13), Barton and co-workers (1996); (14), Wrangham and co-workers (1992); (15), Ron and co-workers (1996); (16), Furuichi (1997); (17), Hausfater (1975); (18), Watts (1994); (19), Horrocks and Hunte (1983); (20), Seyfarth (1976); (22), Fedigan (1993); (23), Mitchell and co-workers (1991); (24), Matsumura (1998); (25), Sterck and Steenbeek (1997); (26), Hill and Okayasu (1995); (27), Cords (2000); (28), Koenig and co-workers (2004).

contests. Such a pattern is somewhat similar to what I report here for patas monkeys in some regard. If dominance is an organizing feature of primate society, species with less stable hierarchies may actually exhibit more agonism than species characterized by a stable, linear hierarchy.

About half of the species for which rates of food-related agonism were reported contested foods at higher rates than both vervets and patas monkeys on Segera. Only chacma baboons and Costa Rican squirrel monkeys exhibited much lower rates of food-related contest competition. Japanese macaques on Koshima Islet, Moor macaques, black spider monkeys, red colobus monkeys, and South American squirrel monkeys exhibited rates of feeding competition similar to those of vervets and patas monkeys on Segera. In terms of the proportion of agonism that feeding contests comprised, Thomas langurs, blue monkeys, and Japanese macaques exhibited proportions of food-related competition as high as vervets on Segera, whereas patas monkeys in this study were similar to Hanuman langurs, chimpanzees, black spider monkeys, and mountain gorillas regarding the proportion of their contests that were associated with food. The discrepancies in patterns of behavior based on what we know so far of primate behavior and ecology in regard to predictions of the models of female social behavior may indicate a greater complexity than simple models have predicted. However, comparisons across sites and studies are lacking for methodological reasons (see Table 7–1).

In order to reliably test predictions regarding dominance relationships among female primates, and to reliably compare species and across sites, standardized operational definitions of variables should be established. The lack of standardization in describing behavior (e.g., what is a stable, linear dominance hierarchy? What type of interactions should be used to determine female social relationships?), making assumptions about social relationships based on complex social interactions (e.g., linking dominance hierarchies to feeding competition without considering other factors), and the variability in measuring and describing food availability discussed in this study are all factors that hinder a proper understanding of the relationships between female primate social behavior and feeding ecology. Isbell and Young (2002) maintain that statistical testing does not correctly interpret the dominance patterns that characterized certain primate species. Instead they propose using latency to detection of a hierarchy. Whereas a linear hierarchy could be detected for vervets after only 410 hours of observation, for patas monkeys a total of 760 hours was necessary to detect a linear hierarchy at the same Segera study site (Isbell & Young 2002).

Although females are expected to contest other resources in addition to food, contests that occur over foods are the basis of current models of female primate social behavior. If a considerable proportion of female contests serve to maintain a hierarchy (i.e., Silk 2003), and food-related competition is not prevalent, the focus should be on contexts in which dominance provides significant benefits to high-ranking individuals. Contests among gorillas, for example, may be related to the importance of other resources such as proximity to males, which has been deemed an important variable to consider by the most recent model of Sterck and co-workers (1997). However, the number of species to which this particular variable is important appears to be limited (Sussman, Garber, & Cheverud 2005). Current socioe-cological models' explanatory power seem especially weak in regard to Strepsirrhine primates (Kappeler 1997) and continue to break down as more quantitative data from primate species become available. Thierry and co-workers (2000) maintain that phylogenetic inertia, rather than eco-logy, can explain many of the features of social behavior and organization exhibited by macaques. The implication therefore is that the behaviors in question must be examined from the point of view of their functional sig-nificance in the past.

## SUMMARY AND IMPLICATIONS FOR FUTURE RESEARCH

Each of the models examined assumes that female primates form social groups to defend resources; as a strategy to reduce the risk of predation; or as a result of some combination of these two factors (Isbell 1991; van Schaik 1989; Sterck, Watts, & van Schaik 1997; Wrangham 1980). Wrangham (1980) and Isbell (1991) hypothesize that primates live in social groups mainly to defend clumped food resources. van Schaik (1989) and Sterck and co-workers (1997) propose that predation is the primary cause of pri-mate social grouping. Of the four models examined here, van Schaik (1989) and Isbell (1991) provide the most specific predictions regarding when and how adult females should compete for food and, consequently, what types of dominance relationships these females will exhibit (e.g., linear and stable). None of the models take into account many of the vari-ables considered important here, although Wrangham's (1980) reference to "high-quality" foods could be interpreted as those that are relatively abundant at feeding sites, widely distributed in space, high in nutritional content, and have low processing costs. According to Wrangham's (1980) model, the stable and linear dominance hierarchy of female vervets indi-cates that they are food limited, but the low degree of feeding competition among vervets on Segera does not support that expectation for monkeys at this site.

Female feeding competition and dominance hierarchies are highly vari-able in a number of species (e.g., guenons, colobines, and lemurids), which may preclude the construction of a simple model to explain female social behavior for all primate species. Among adult female vervets on Segera, both habitat and food-related contests affected the linearity of dominance hierarchies. However, foods that were contested did not conform to the gen-eral patterns of food availability predicted by models of female social behavior, and vervets contested other resources besides foods. "Habitat" was a misleading indicator of competition because the same type of habitat used by patas monkeys and vervets differed significantly in some respects between the home ranges of the two species. Additionally, among patas monkeys on Segera, other factors besides food-related contests were equally important in influencing female social relationships. The results of this study demonstrate the variability of female primate social behavior and the need to account for multiple factors that may influence patterns of female social behavior.

The categories used by models of female primate social behavior to denote patterns of food availability are too broad, and they fail to distinguish between the distribution and abundance of foods or, except for Wrangham (1980), to indicate food quality. For example, van Schaik (1989) predicts that foods that are dispersed or distributed in clumps greater than group size promote only strong within-group scramble competition. The term "dispersed" does not take into account the abundance of all foods. For example, a few, small, dispersed patches are likely to promote stronger contest competition than many small, dispersed patches.

A variable not included in the models that may be relevant to adult female patas monkeys' sociality is female tenure in a social group, a function of demography. In comparison with species such as savanna baboons, stable female dominance relations among patas monkeys would not be expected. Olive baboon females first give birth, on average, when they are almost 7 years old (Bercovitch & Strum 1993). Patas monkeys first give birth at 3 years of age and then usually during each subsequent year (Chism & Rowell 1988). The nepotistic nature of dominance relationships among female baboons and macaques is thought to contribute to the stability of female hierarchies in these species. Adult females assist their daughters in attaining rank immediately below them in the hierarchy, and subsequent daughters usually move ahead of their older sisters to assume rank below their mother. If female patas monkeys' tenure in a group were short, maternal assistance in daughter's rank acquisition would be uncommon, especially given the high mortality of patas monkeys in Laikipia (Chism & Rowell 1988; Enstam & Isbell 2002). The lack of alliance behavior in patas monkeys on Segera supports this argument. Additionally, the rapid maturation of female patas monkeys is not conducive to strong mother–infant bonds compared to species that mature more slowly (e.g., baboons and macaques) (Chism 1986).

The apparent contradiction between the low rate of contest competition exhibited by adult female vervets and their stable, linear dominance hierarchy illustrates the importance of considering behavior at the proximate level (see also Matsumura 1998). For example, contest competition among females over food reflects the influence of ecology on behavior. However, contests over other resources, such as access to mates, may also influence the type of dominance relationships exhibited by adult females in a group. Approximately half of all agonistic interactions among adult female patas monkeys were not food related. Dominance relationships are not only influenced by access to food resources and, thus, models of female social relationships might be more accurate if variables such as those related to demography (as in the case of patas monkeys) or other variables are considered.

Barton and co-workers (1996) constructed a model of ecology, feeding competition, and social structure for baboon species based on the model proposed by van Schaik (1989) but included the influence of male–male competition on baboon social structure. Where foods are clumped and predation is high, within-group competition is expected to be high, and social groups will consequently be large and "female bonded" (Barton, Byrne, & Whiten 1996). Vervet behavior on Segera corresponded to the predictions set forth by the baboon model only if foods within the riverine habitat are considered to characterize the foods available to vervets on Segera. However, I demonstrated that the relationships between female dominance behavior and food availability are misleading. Where foods are dispersed and predation pressure is high, within-group competition is expected to be lower, and social groups are expected to be large but composed of smaller, non-female bonded and often single-male units (Barton, Byrne, & Whiten 1996). Patas monkeys on Segera conform to the latter prediction, with the exception of female residency. The fact that patas monkey social groups contain only a single resident male may preclude the occurrence of small, non-female bonded units because group fissioning among baboons is attributed to male–male competition.

Several studies demonstrate clear benefits to dominant individuals during periods of resource scarcity (Cheney, Lee, & Seyfarth 1981; Wrangham 1981). Perhaps a new generation of models (Isbell & Young 2002) could consider those resources most limiting to female primates in such a context because this may be the most critical time for a female's reproductive success. Keystone resource (Terborgh 1983) or fallback food distribution and abundance may prove to be a better determinant of female primate social organization than the sum of all foods in the diet. Because female primates that exhibit stable, linear dominance hierarchies contest a number of resources in addition to food, as well as engage in "random" acts of aggression to assert their dominance (Silk 2002), a more specific model should consider food availability as a selective pressure at times when resources are most scarce or during critical periods (Wrangham 1980). Taking into consideration more than twenty ecological variables, Stevenson (2004) found that woolly monkeys in Colombia preferred fruits from large trees that were in fruit during times of fruit scarcity. Cheney and co-workers (1981) demonstrated that low-ranking female vervets exhibited a significantly higher degree of mortality due to disease apparently due to restricted access to food and water at Amboseli, Kenya. Similarly, Wrangham (1981) found low-ranking females at this same site were excluded from water sources during drought periods. As an organizing aspect of primate society, female dominance hierarchies might be expected to reflect more accurately the distribution and abundance of keystone resources or fallback foods. For primates living in arid environments, such as some baboons and chimpanzees, vervets, and patas monkeys, the underlying social structure may reflect water availability during the dry season.

Ultimately, long-term data on female reproductive success are necessary to establish the effects of rank on females and to elaborate on the resources that are most crucial to females' reproductive success. Among Gombe chimpanzees, female dominance was ascertained only after years of study but was found to confer significant benefits to high-ranking females (Pusey, Williams, & Goodall 1997). The evolutionary explanation for the existence of dominance hierarchies within social groups is often cast in terms of access to preferred resources. Discovering what these resources are will be the key to predicting patterns of female social behavior. As more data become available, using gross categories of diet will likely prove to be too generalized a classification to predict female primate social behavior (see also Koenig 2002). As Whitten (1983) demonstrated for vervets and as I have shown here, specific foods important in the diet are those that should be considered first, and further studies that relate dominance rank to reproductive success may show that keystone resources or fallback foods are also good indicators of female social behavior in primates.

A possible explanation for the maintenance of stable, linear dominance hierarchies among female vervets in the absence of limited food availability is the assurance of priority of access to resources for dominant individuals during periods of resource scarcity. Rowell (1972) suggested that clear-cut hierarchies themselves might be viewed as a stress symptom, in that low-ranking individuals consistently exhibit chronic, intense, subordinate behavior. Following this, a dominance hierarchy would represent ". . . a sort of a pathological response of a social system to extremely stressful environmental conditions." (Rowell 1972: 163). In species that are characteristically found in areas of environmental extremes (e.g., severe dry season droughts common), insurance against limited access to resources would be adaptive. In the event that resources became so limiting that certain individuals would suffer, being of high rank would ensure access to resources with less cost related to aggression than if dominance relationships were undecided.

In summary, models that have been proposed to explain female social relationships among primates have categorized patterns of food availability such that they are of limited value in predicting female social relationships among most primate species. Additionally, researchers who test hypotheses related to questions of female primate social behavior and food availability rarely quantify all relevant variables adequately. My findings indicate that models of female primate social behavior are currently too broad to accurately describe conditions of food availability predicted to lead to contest competition (Figure 7–1). Whereas each of the models focuses on food distribution, in my study, other variables such as a food's abundance

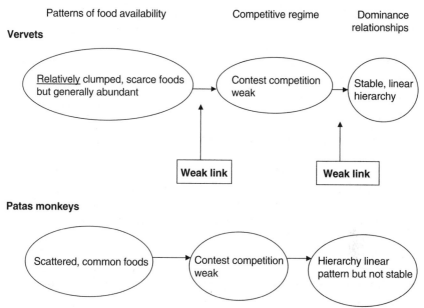

**Figure 7–1** Food availability, feeding competition, and dominance relationships in free-ranging vervets and patas monkeys in Laikipia, Kenya.

and relative importance in the diet were more influential in determining when primates would contest for foods. Additionally, variables besides food resource availability had to be considered in order to fully understand the nature of female vervet and patas monkey competition and dominance; in cases in which food resources did influence competition and dominance among primates in this study, the current terminology used to describe food resource availability in primates is much too vague.

In order to adequately test the predictions proposed by Isbell (1991), van Schaik (1989), Sterck and co-workers (1997), and Wrangham (1980), studies of primate behavior must (1) adequately assess an array of factors that influence primates' perception of their available food resources (i.e., abundance, distribution, nutrient content, processing costs, and availability of other foods), (2) reliably measure the relevant components affecting female social relationships (e.g., contest competition as well as particular styles of dominance relationships), and (3) take into consideration what other factors besides feeding competition may influence adult female primate social relationships (e.g., competition over safe positions within a group and access to grooming partners). The results of this research demonstrate the dangers of assuming relationships between female primate social behavior and ecology without quantifying the appropriate variables necessary to test related hypotheses.

# References

*Afr J Ecol*. Notes and records: Population characteristics of the vervet monkey in the Mosi-Oa-Tunya National Park, Zambia. 32: 72–74.

Agetsuma, N. 1995a. Foraging strategies of Yakushima macaques (*Macaca fuscata yakui*). *Int J Primatol* 16: 595–609.

————. 1995b. Dietary selection by Yakushima macaques (*Macaca fuscata yakui*): The influence of food availability and temperature. *Int J Primatol* 16: 611–27.

Altmann, J. 1974. Observational study of behaviour: Sampling methods. *Behaviour* 49: 227–65.

————. 1980. *Baboon mothers and infants*. Cambridge, MA: Harvard University Press.

Altmann, S. A., D. G. Post, and D. F. Klein. 1987. Nutrients and toxins of plants in Amboseli, Kenya. *Afr J Ecol* 25: 279–93.

Appleby, M. C. 1983. The probability of linearity in hierarchies. *Anim Behav* 31: 600–8.

Barrette, C., and D. Vandal. 1986. Social rank, dominance, antler size, and access to food in snowbound wild woodland caribou. *Behaviour* 97: 118–46.

Barton, R. A. 1990. Feeding, reproduction and social organisation in female olive baboons (*Papio anubis*). In *Baboons: Behavior and ecology, use and care*, ed. M. T. de Mello, A. Whiten, and R. W. Byrne, 29–37. Brasil: Selected Proceedings of the XII Congress of the International Primatological Society.

————. 1993. Sociospatial mechanisms of feeding competition in female olive baboons, *Papio anubis*. *Anim Behav* 46 (4): 685–94.

Barton, R. A., and A. Whiten. 1993. Feeding competition among female olive baboons, *Papio anubis*. *Anim Behav* 46: 777–89.

Barton, R. A., R. W. Byrne, and A. Whiten. 1996. Ecology, feeding competition and social structure in baboons. *Behav Ecol Sociobiol* 38: 321–9.

Barton, R. A., A. Whiten, R. W. Byrne, and M. English. 1993. Chemical composition of baboon plant foods: Implications for the interpretation of intra- and interspecific differences in diet. *Folia Primatol* 61: 1–20.

Barton, R. A., A. Whiten, S. C. Strum, R. W. Byrne, and A. J. Simpson. 1992. Habitat use and resource availability in baboons. *Anim Behav* 43: 831–44.

Bercovitch, F. B., and S. C. Strum. 1993. Dominance rank, resource availability, and reproductive maturation in female savanna baboons. *Behav Ecol Sociobiol* 33: 313–18.

Bernstein, I. S. 1981. Dominance: The baby and the bathwater. *Behav Brain Sci* 4: 419–57.

Bernstein, I. S., and C. L. Ehardt. 1985. Intragroup agonistic behavior in rhesus monkeys (*Macaca mulatta*). *Int J Primatol* 6: 209–26.

Boesch, C. 1996. Social grouping in Tai chimpanzees. In *Great ape societies*, ed. W. C. McGrew, L. F. Marchant, and T. Nishida, 101–13. Great Britain: Cambridge University Press.

Borries, C. 1993. Ecology of female social relationships: Hanuman langurs (*Presbytis entellus*) and the van Schaik model. *Folia Primatol* 61: 21–30.

Budnitz, N. 1978. Feeding behavior of *Lemur catta* in different habitats. In *Perspectives in ethology*. Vol. 3, *Social behavior*, ed. P. P. G. Bateson and P. H. Klopfer, 85–108. New York: Plenum Press.

Butynski, T. 1988. Guenon birth seasons and correlates with rainfall and food. In *A primate radiation: Evolutionary biology of the African guenons*, ed. A. Gautier-Hion, F. Bourliere, J. Gautier, and J. Kingdon, 284–322. Cambridge: Cambridge University Press.

Bygott, J. D. 1979. Agonistic behavior, dominance, and social structure in wild chimpanzees of the Gombe National Park. In *The great apes*, ed. D. A. Hamburg and E. R. McGown, 405–27. Menlo Park, CA: The Benjamin/Cummings Publishing Company, Inc.

Byrne, R. W., A. Whiten, and S. P. Henzi. 1990. Measuring the food constraints of mountain baboons. In *Baboons: Behaviour and ecology, use and care*, ed. M. T. de Mello, A. Whiten, and R. W. Byrne, 105–22. Brasil: Selected Proceedings of the XII Congress of the International Primatological Society.

Carlson, A. A., and L. A. Isbell. 2001. Causes and consequences of single-male and multimale mating in free-ranging patas monkeys, *Erythrocebus patas*. *Anim Behav* 62: 1047–58.

Chapman, C. 1985. The influence of habitat on behaviour in a group of St. Kitts green monkeys. *J Zool Lond* 206: 311–20.

———. 1988. Patch use and patch depletion by the spider and howling monkeys of Santa Rosa National Park, Costa Rica. *Behaviour* 105: 99–116.

Chapman, C. A., and L. J. Chapman. 2000. Determinants of group size in primates: The importance of travel costs. In *On the move: How and why animals travel in groups*, ed. S. Boinski and P. A. Garber, 24–42. Chicago: Chicago University Press.

Chapman, C., F. J. White, and R. Wrangham. 1994. Party size in chimpanzees and bonobos. In *Chimpanzee cultures*, ed. R. W. Wrangham, W. C. McGrew, F. B. M. deWaal, and P. Heltne, 41–57. Cambridge, MA: Harvard University Press.

Chapman, C. A., L. Chapman, K. A. Bjorndal, and D. A. Onderdonk. 2002. Application of protein-to-fiber ratios to predict Colobine abundance on different spatial scales. *Int J Primatol* 23: 283–310.

Chapman, C. A., L. J. Chapman, L. Naughton-Treves, M. J. Lawes, L. R. McDowell. 2004. Predicting folivorous primate abundance: Validation of a nutritional model. *Am J Primatol* 62: 55–69.

Chapman, C. A., L. J. Chapman, K. D. Rode, E. M. Hauck, and L. R. McDowell. 2003. Variation in the nutritional value of primate foods: Among trees, time periods, and areas. *Int J Primatol* 24: 317–33.

Chapman, C., L. J. Chapman, R. W. Wrangham, K. Hunt, D. Gebo, and L. Gardner. 1992. Estimators of fruit abundance of tropical trees. *Biotropica* 24: 527–31.

Cheney, D. L., and R. M. Seyfarth. 1983. Non-random dispersal in free-ranging vervet monkeys: Social and genetic consequences. *Am Nat* 122: 392–412.

———. 1987. The influence of intergroup competition on the survival and reproduction of female vervet monkeys. *Behav Ecol Sociobiol* 21: 375–86.

———. 1990. *How monkeys see the world*. Chicago: University of Chicago Press.

Cheney, D. L., and R. W. Wrangham. 1987. Predation. In *Primate societies*, ed. B. B. Smuts, D. L. Cheney, R. M. Seyfarth, R. W. Wrangham, and T. T. Struhsaker, 227–39. Chicago: University of Chicago Press.

Cheney, D. L., P. C. Lee, and R. M. Seyfarth. 1981. Behavioral correlates of non-random mortality among free-ranging female vervet monkeys. *Behav Ecol Sociobiol* 9: 153–61.

Cheney, D. L., R. M. Seyfarth, J. Fischer, J. Beehner, T. Bergman, S. E. Johnson, D. M. Kitchen, R. A. Palombit, D. Rendall, and J. B. Silk. 2004. Factors affecting reproduction and mortality among baboons in the Okavango Delta, Botswana. *Int J Primatol* 25: 401–28.

Chism, J. 1986. Development and mother-infant relationships among captive patas monkeys. *Int J Primatol* 7: 49–81.

Chism, J., and T. Rowell. 1986. Mating and residence patterns of male patas monkeys. *Ethology* 72: 31–9.

———. 1988. The natural history of patas monkeys. In *A primate radiation: Evolutionary biology of the African guenons*, ed. A. Gautier-Hion, F. Bourliere, J. Gautier, and J. Kingdon, 412–38. Cambridge: Cambridge University Press.

Chism, J., and C. S. Wood. 1994. Diet and feeding behavior of patas monkeys (*Erythrocebus patas*) in Kenya. *Am J Phys Anthropol Suppl* 18: 67.

Chism, J., T. Rowell, and D. K. Olson. 1984. Life history patterns of female patas monkeys. In *Female primates: Studies by women primatologists*, ed. M. Small, 175–92. New York: Alan R. Liss.

Clarke, R. 1986. *The handbook of ecological monitoring.* Oxford: Clarendon Press.

Clemmons, J. R., and R. Buchholz. 1997. Linking conservation and behavior. In *Behavioral approaches to conservation in the wild,* ed. J. R. Clemmons and R. Buchholz, 3–22. Cambridge: Cambridge University Press.

Clutton-Brock, T. H., and P. H. Harvey. 1976. Species differences in feeding and ranging behavior in primates. In *Primate ecology: Studies of feeding and ranging behaviour in lemurs, monkeys, and apes,* 539–56. London: Academic Press.

Coe, M., and H. Beentje. 1991. *A field guide to the acacias of Kenya.* Oxford: Oxford University Press.

Collins, D. A., and W. C. McGrew. 1988. Habitats of three groups of chimpanzees (*Pan troglodytes*) in western Tanzania compared. *J Hum Evol* 17 (6): 553–74.

Cords, M. 1988. Mating systems of forest guenons: A preliminary view. In *A primate radiation: Evolutionary biology of the African guenons,* ed. A. Gautier-Hion, F. Bourliere, J. Gautier, and J. Kingdon, 323–39. Cambridge: Cambridge University Press.

Cords, M. 2000. The number of males in guenon groups. In *Primate males: Causes and consequences of variation in group composition,* ed. P. M. Kappeler, 84–96. Cambridge: Cambridge University Press.

Corsi, F., J. de Leeuw, and A. K. Skidmore. 2000. Modeling species distribution with GIS. In *Research techniques in animal ecology: Controversies and consequences,* ed. L. Boitani and T. K. Fuller, 389–434. New York: Columbia University Press.

Cottam, G., and J. T. Curtis. 1956. The use of distance measures in phytosociological sampling. *Ecology* 37: 451–61.

Crook, J. H., and J. S. Gartlan. 1966. Evolution of primate societies. *Nature* 210: 1200–3.

Dagg, A. I., and J. B. Foster. 1976. *The giraffe: Its biology, behavior and ecology.* Malabar, FL: Robert E. Krieger Publishing Company.

Disotell, T. 1996. Phylogeny of Old World monkeys. *Evol Anthropol* 5 (1): 18–24.

———. 2000. Molecular systematics of the Cercopithecidae. In *Old World monkeys,* ed. P. F. Whitehead and C. J. Jolly, 29–56. Cambridge: Cambridge University Press.

Doran, D. 1997. Influence of seasonality on activity patterns, feeding behavior, ranging, and grouping patterns in Tai chimpanzees. *Int J Primatol* 18 (2): 183–206.

Doran, D., and A. McNeilage. 1998. The density of herbaceous vegetation in south-west Central African Republic: Implications for western lowland gorilla (*Gorilla gorilla gorilla*) socioecology. Abstract. *Am J Phys Anthropol* 26: 77–78.

Dougall, H. W., V. M. Drysdale, and P. E. Glover. 1964. The chemical composition of Kenya browse and pasture herbage. *East Afr Wild J* 2: 82–125.

Drews, C. 1993. The concept and definition of dominance in animal behavior. *Behaviour* 125 (3–4): 283–313.

Drickamer, L. 1974. A ten-year summary of reproductive data for free-ranging *Macaca mulatta*. *Folia Primatol* 21: 61–80.

Dunbar, R. I. M. 1988. *Primate social systems*. New York: Comstock Publishing Associates.

Enstam, K. L., and L. A. Isbell. 2002. Comparison of responses to alarm calls by patas (*Erythrocebus patas*) and vervet (*Cercopithecus aethiops*) monkeys in relation to habitat structure. *Am J Phys Anthropol* 119: 3–14.

Enstam, K. L., L. A. Isbell, and T. W. De Maar. 2002. Male demography, female mating behavior, and infanticide in wild patas monkeys (*Erythrocebus patas*). *Int J Primatol* 23: 85–104.

Fedigan, L. 1993. Sex differences and intersexual relations in adult white-faced capuchins (*Cebus capucinus*). *Int J Primatol* 14 (6): 853–77.

Fedigan, L., and L. M. Fedigan. 1988. *Cercopithecus aethiops*: A review of field studies. In *A primate radiation: Evolutionary biology of the African guenons*, ed. A. Gautier-Hion, F. Bourliere, J. Gautier, and J. Kingdon, 389–411. Cambridge: Cambridge University Press.

Fossey, D., and A. H. Harcourt. 1977. Feeding ecology of free-ranging Mountain Gorilla (*Gorilla gorilla beringei*). In *Primate ecology. Studies of feeding and ranging behaviour in lemurs, monkeys and apes*, ed. T. H. Clutton-Brock, 415–47. London: Academic Press.

Furuichi, T. 1997. Agonistic interactions and matrifocal dominance rank of wild bonobos (*Pan paniscus*) at Wamba. *Int J Primatol* 18 (6): 855–75.

Garber, P. A. 1986. The ecology of seed dispersal in two species of Callitrichid primates (*Saguinus mystax* and *Saguinus fuscicollis*). *Am J Primatol* 10: 155–70.

———. 1988. Diet, foraging patterns, and resource defense in a mixed species troop of *Saguinus mystax* and *Saguinus fuscicollis* in Amazonian Peru. *Behaviour* 105: 18–34.

———. 1997. Primate behavioral ecology. In *Encyclopedia of human biology*, 85–92. 2nd ed. Vol. 7. London: Academic Press.

Garland, T., and S. C. Adolph. 1994. Why not to do two-species comparative studies: Limitations on inferring adaptation. *Physiol Zool* 67 (4): 797–828.

Gartlan, J. S., and C. K. Brain. 1968. Ecology and social variability in *Cercopithecus aethiops* and *C. mitis*. In *Primates; studies in adaptation and variability*, ed. P. C. Jay, 253–92. New York: Holt, Rinehart & Winston.

Gautier, J. 1988. Interspecific affinities among guenons as deduced from vocalizations. In *A primate radiation: Evolutionary biology of the African guenons*, ed. A. Gautier-Hion, F. Bouliere, J. Gautier, and J. Kingdon, 194–226. Cambridge: Cambridge University Press.

Glander, K. E. 1978. Howling monkey feeding behaviour and plant secondary compounds: A study of strategies. In *The ecology of arboreal folivores*, ed. G. C. Montgomery. Washington, DC: Smithsonian University Press.

Goldizen, A. W., J. Terborgh, F. Cornejo, D. T. Porras, and R. Evans. 1988. Season food shortage, weight loss, and the timing of births in saddle-back tamarins (*Saguinus fuscicollis*). *J Anim Ecol* 57: 893–901.

Goldman, E. N., and J. Loy. 1997. Longitudinal study of dominance relations among captive patas monkeys. *Am J Primatol* 42: 41–51.

Goodall, J. 1986. *The chimpanzees of Gombe: Patterns of behaviour*. Cambridge, MA: Belknap Press.

Gore, M. A. 1993. Effects of food distribution on foraging competition in rhesus monkeys, *Macaca mulatta*, and hamadryas baboons, *Papio hamadryas*. *Anim Behav* 45: 773–86.

Gouzoules, S., and H. Gouzoules. 1987. Kinship. In *Primate Societies*, ed. B. B. Smuts, D. L. Cheney, R. M. Seyfarth, R. W. Wrangham, and T. T. Struhsaker, 299–305. Chicago: University of Chicago Press.

Groves, C. 2001. *Primate taxonomy*. Washington, DC: Smithsonian Institution Press, 350 pp.

Guillotin, M., G. Dubost, and D. Sabatier. 1994. Food choice and food competition among the three major primate species of French Guiana. *J Zool Lond* 233: 551–79.

Hall, K. R. L. 1965. Ecology and behavior of baboons, patas, and vervet monkeys in Uganda. In *The baboon in medical research*, ed. H. Vagtborg, 43–61. Austin: University of Texas Press.

———. 1967. Social interactions of the adult male and adult females of a patas monkey group. In *Social communication among primates*, ed. S. A. Altmann, 261–80. Chicago: University of Chicago Press.

Hall, K. R. L., and J. S. Gartlan. 1965. Ecology and behaviour of the vervet monkey, *Cercopithecus aethiops*, Lolui Island, Lake Victoria. *Proc Zool Soc Lond* 145: 37–57.

Haltenorth, T., and H. Diller. 1977. *A field guide to the mammals of Africa including Madagascar*. London: Collins Sons & Co Ltd.

Hamilton III, W. J., R. E. Buskirk, and W. H. Buskirk. 1976. Defense of space and resources by chacma (*Papio ursinus*) baboon troops in an African desert and swamp. *Ecology* 57: 1264–72.

Harcourt, A. H. 1987. Dominance and fertility among female primates. *J Zool Lond* 213: 471–87.

Harrison, M. J. S. 1983. Age and sex differences in the diet and feeding strategies of the green monkey, *Cercopithecus sabaeus*. *Anim Behav* 31: 969–77.

———. 1984. Optimal foraging strategies in the diet of the green monkey, *Cercopithecus sabaeus*, at Mt. Assirik, Senegal. *Int J Primatol* 5 (5): 435–71.

———. 1985. Time budget of the green monkey, *Cercopithecus sabaeus*: Some optimal strategies. *Int J Primatol* 6 (4): 351–76.

Hausfater, G. 1975. *Dominance and reproduction in baboons (Papio cynocephalus)*. Vol. 7, *Contributions to Primatology*. Basel: S. Karger.

Henzi, S. P., and J. W. Lucas. 1980. Observations on the inter-troup movement of adult vervet monkeys (*Cercopithecus aethiops*). *Folia Primatol* 43: 189–97.

Henzi, S. P., J. E. Lycett, and S. E. Piper. 1997. Fission and troop size in a mountain baboon population. *Anim Behav* 53 (3): 525–35.

Hill, D.A., and N. Okayasu. 1995. Absence of youngest ascendance in the dominance relations of sisters in wild Japanese macaques (*Macaca fuscata yakui*). *Behaviour* 132: 367–79.

Hinde, R. A. 1979. The nature of social structure. In *The great apes*, ed. D. A. Hamburg and E. R. McCown, 295–315. Menlo Park, CA: The Benjamin/ Cummings Publishing Company, Inc.

Hocking, B. 1970. Insect associations with the swollen thorn acacias. *Trans Rent Soc Lond* 122 (7): 211–55.

van Hoof, J. A. R. A. M. 1974. A structural analysis of the social behavior of a semi-captive group of chimpanzees. In *Social communication and movement*, ed. M. Cranach and I. Vine, 75–162. London: Academic Press.

Horrocks, J. A., and W. Hunte. 1983. Maternal rank and offspring rank in vervet monkeys: An appraisal of the mechanisms of rank acquisition. *Anim Behav* 31: 772–82.

———. 1993. Interactions between juveniles and adult males in vervets: Implications for adult male turnover. In *Juvenile primates*, ed. M. E. Pereira and L. A. Fairbanks, 228–39. New York: Oxford University Press.

Hrdy, S. 1977. *The langurs of Abu. Female and male strategies of reproduction.* Cambridge, MA: Harvard University Press.

Hurov, J. R. 1987. Terrestrial locomotion and back anatomy in vervets (*Cercopithecus aethiops*) and patas monkeys (*Erythrocebus patas*). *Am J Primatol* 13: 297–311.

Ihobe, H. 1989. How social relationships influence a monkey's choice of feeding sites in the troop of Japanese macaques (*Macaca fuscata fuscata*) on Koshima Islet. *Primates* 30 (1): 17–25.

Isabirye-Basuta, G. 1988. Food competition among individuals in a free-ranging chimpanzee community in Kibale Forest, Uganda. *Behaviour* 105: 135–47.

Isbell, L. A. 1991. Contest and scramble competition: Patterns of female aggression and ranging behavior among primates. *Behav Ecol* 2: 143–55.

———. 1998. Diet for a small primate: Insectivory and gummivory in the (large) patas monkey (*Erythrocebus patas pyrrhonotus*). *Amer J Primatol* 45: 381–98.

Isbell, L. A., D. L. Cheney, and R. M. Seyfarth. 1990. Costs and benefits of home range shifts among vervet monkeys (*Cercopithecus aethiops*) in Amboseli National Park, Kenya. *Behav Ecol Sociobiol* 27: 351–58.

Isbell, L. A., and K. L. Enstam. 2002. Predator (in)sensitive foraging in sympatric female vervets (*Cercopithecus aethiops*) and patas monkeys (*Erythrocebus patas*): A test of ecological models of group dispersion. In *Eat or be eaten: Predator sensitive foraging among primates*, ed. L. E. Miller, 154–68. Cambridge: Cambridge University Press.

Isbell, L. A., and J. D. Pruetz. 1998. Differences between patas monkeys (*Erythrocebus patas*) and vervet monkeys (*Cercopithecus aethiops*) in agonistic interactions between adult females. *Int J Primatol* 19 (5): 837–55.

Isbell, L. A., and D. van Vuren. 1996. Differential costs of locational and social dispersal and their consequences for female group-living primates. *Behaviour* 133: 1–29.

Isbell, L. A., and T. P. Young. 2002. Ecological models of female social relationships in primates: Similarities, disparities, and some directions for future clarity. *Behaviour* 139: 177–202.

Isbell, L. A., J. D. Pruetz, and T. P. Young. 1998. Movements of vervets (*Cercopithecus aethiops*) and patas monkeys (*Erythrocebus patas*) as estimators of food resource size, density, and distribution. *Behav Ecol Sociobiol* 42: 123–33.

Isbell, L. A., J. D. Pruetz, T. P. Young, and M. Lewis. 1998. Locomotory activity differences between sympatric patas (*Erythrocebus patas*) and vervet monkeys (*Cercopithecus aethiops*): Implications for the evolution of long hindlimb length in *Homo*. *Am J Phys Anthropol* 105 (2): 199–207.

Iwamoto, T. 1992. Range use patterns in relation to resource distribution of free-ranging Japanese monkeys. In *Topics in Primatology*. Vol. 2, *Behavior, ecology and conservation*, ed. N. Itoigawa, Y. Sugiyama, G. P. Sackett, and R. K. R. Thompson, 57–65. Tokyo: University of Tokyo Press.

Izar, P. 2004. Female social relationships of *Cebus apella nigritus* in a southeastern Atlantic forest: An analysis through ecological models of primate social evolution. *Behaviour* 141: 71–99.

Janson, C. H. 1985. Aggressive competition and individual food consumption in wild brown capuchin monkeys (*Cebus apella*). *Behav Ecol Sociobiol* 18: 125–38.

———. 1988. Food competition in brown capuchin monkeys (*C. apella*): Quantitative effects of group size and tree productivity. *Behaviour* 105: 53–75.

Janson, C. H., and C. A. Chapman. 1999. Resources and primate community structure. In *Primate communities*, ed. J. G. Fleagle, C. Janson, and K. E. Reed, 237–67. Cambridge: Cambridge Universtiy Press.

Janson, C. H., and M. Goldsmith. 1995. Predicting group size in primates: Foraging costs and predation risks. *Behav Ecol* 6: 326–36.

Johnson, J. 1989. Supplanting by olive baboons: Dominance rank difference and resource value. *Behav Ecol Sociobiol* 24: 277–83.

Johnson, R. B. 1990. The effects of plant chemistry on food selection by adult male yellow baboons (*Papio cynocephalus*). In *Baboons: Behavior and ecology, use and care*, ed. M. T. de Mello, A. Whiten, and R. W. Byrne, 23–28. Brasil: Selected Proceedings of the XII Congress of the International Primatological Society.

Kaplan, J. R. 1987. Dominance and affiliation in the Cercopithecini and Papionini: A comparative examination. In *Comparative behavior of African monkeys*, ed. E. L. Zucker, 127–50. New York: Alan R. Liss Inc.

Kaplan, J. R., and E. Zucker. 1980. Social organization in a group of free-ranging patas monkeys. *Folia Primatol* 34: 196–213.

Kappeler, P. M. 1997. Intrasexual selection in Mirza coquereli: Evidence for scramble competition polygyny in a solitary primate. *Behav Ecol Sociobiol* 41: 115–27.

Kavanagh, M. 1978. The diet and feeding behavior of *Cercopithecus aethiops tantalus*. *Folia Primatol* 30: 30–63.

Kendall, M. G. 1962. *Rank correlation methods*. London: Charles Griffin.

Kerr, K. B. 1998. Variation in reproduction of female Gombe chimpanzees. Abstract. *Am J Primatol* 45 (2): 190.

Kingdon, J. 1988. Comparative morphology of hands and feet in the genus *Cercopithecus*. In *A primate radiation: Evolutionary biology of the African guenons*, ed. A. Gautier-Hion, F. Bourliere, J. Gautier, and J. Kingdon, 184–93. Cambridge: Cambridge University Press.

Klein, D. F. 1978. The diet and reproductive cycle of a population of vervet monkeys (*Cercopithecus aethiops*). Ph.D. diss., New York University.

Koenig, A. 2002. Competition for resources and its behavioral consequences among female primates. *Int J Primatol* 23: 759–83.

Koenig, A., J. Beise, M. K. Chalise, and J. U. Ganzhorn. 1998. When females should contest for food-testing hypotheses about resource density, distribution, size, and quality with Hanuman langurs (*Presbytis entellus*). *Behav Ecol Sociobiol* 42: 225–37.

Koenig, A., E. Larney, A. Lu, and C. Borries. 2004. Agonistic behavior and dominance relationships in female Phayre's leaf monkeys–preliminary results. *Am J Primatol* 64: 351–7.

Korstjens, A. H. 2001. The mob, the secret sorority, and the phantoms: An analysis of the socio-ecological strategies of the three colobines of Tai. Ph.D. diss., Utrecht University, 173 pp.

Kramer, M., and J. Schmidhammer. 1992. The chi-squared statistic in ethology: Use and misuse. *Anim Behav* 44: 833–41.

Kummer, H. 1990. The social system of Hamadryas baboons and its presumable evolution. In *Baboons: Behaviour and ecology, use and care*, ed. M. T. de Mello, A. Whiten, and R. W. Byrne, 43–60. Brasil: Selected Proceedings of the XII Congress of the International Primatological Society.

Kuroda, S., and C. E. G. Tutin. 1993. Field studies of African apes in tropical rain forests: Methods to increase the scope and accuracy of intersite comparisons. *Tropics* 2 (4): 187–8.

van der Kuyl, A. C., C. L. Kuiken, J. T. Dekker, and J. Goudsmit. 1995. Phylogeny of African monkeys based upon mitochondrial 12S rRNA sequences. *J Mol Evol* 40: 173–80.

Leakey, M. 1988. Fossil evidence for the evolution of the guenons. In *A primate radiation: Evolutionary biology of the African guenons*, ed. A. Gautier-Hion, F. Bourliere, J. Gautier, and J. Kingdon, 7–12. Cambridge: Cambridge University Press.

Lee, P. C. 1984. Ecological constraints on the social development of vervet monkeys. *Behaviour* 91: 245–62.

Lernould, J. 1988. Classification and geographical distribution of guenons: A review. In *A primate radiation: Evolutionary biology of the African guenons*, ed. A. Gautier-Hion, F. Bourliere, J. Gautier, and J. Kingdon, 54–78. Cambridge: Cambridge University Press.

Lockwood, R. 1979. Dominance in wolves: Useful construct or bad habit? In *The behaviour and ecology of wolves*, ed. E. Klinghammer, 225–43. New York: Garland STPM Press.

Loy, J., and M. Harnois. 1988. An assessment of dominance and kinship among patas monkeys. *Primates* 29: 331–42.

Loy, J., B. Argo, G. Nestell, S. Vallett, and G. Wanamaker. 1993. A reanalysis of patas monkeys' "grimace and gecker" display and a discussion of their lack of formal dominance. *Int J Primatol* 14 (6): 879–93.

Ludwig, J. A., and J. F. Reynolds. 1988. *Statistical ecology*. New York: John Wiley & Sons.

Madden, D., and T. P. Young. 1992. Ants as alternative defenses in spinescent *Acacia drepanolobium*. *Oecologia* 91: 235–8.

Madrigal, L. 1998. *Statistics for anthropology*. Cambridge: Cambridge University Press, 238 pp.

Malenky, R. 1990. Ecological factors affecting food choice and social organization in *Pan paniscus*. Ph.D. diss., State University of New York at Stony Brook, 302 pp.

Malenky, R. K., R. Wrangham, C. A. Chapman, and E. O. Vineberg. 1993. Measuring chimpanzee food abundance. *Tropics* 2 (4): 231–44.

Marsh, F. 1993. The behavioral ecology of young baboons. Ph.D. thesis, University of St. Andrews, Scotland.

Martin, R. D., and A. M. MacLarnon. 1988. Quantitative comparisons or the skull and teeth in guenons. In *A primate radiation: Evolutionary biology of the African guenons*, ed. A. Gautier-Hion, F. Bourliere, J. Gautier, and J. Kingdon, 160–83. Cambridge: Cambridge University Press.

Mathy, J. W., and L. A. Isbell. 2001. The relative importance of size of food and interfood distance in eliciting aggression in captive rhesus macaques (*Macaca mulatta*). *Folia Primatol* 72: 268–77.

Matsumoto-Oda, A., K. Hosaka, M. A. Huffman, and K. Kawanaka. 1998. Factors affecting party size in chimpanzees of the Mahale mountains. *Int J Primatol* 19: 999–1011.

Matsumura, S. 1998. Relaxed dominance relations among female Moor macaques (*Macaca maurus*) in their natural habitat, South Sulawesi, Indonesia. *Folia Primatol* 69: 346–56.

Matsubayashi, K., M. Hirai, T. Watanabe, Y. Ohkura, and K. Nozawa. 1978. A case of vervet-patas hybrid in captivity. *Primates* 19: 785–93.

McGrew, W. C. 1992. *Chimpanzee material culture. Implications for human evolution*. Cambridge: Cambridge University Press.

Melnick, D. J., and M. C. Pearl. 1987. Cercopithecines in multimale groups: Genetic diversity and population structure. In *Primate societies*, eds. B. B. Smuts, D. L. Cheney, R. M. Seyfarth, R. W. Wrangham, and T. T. Struhsaker, 121–34. Chicago and London: The University of Chicago Press.

Miller, K. E., and J. M. Dietz. 2004. Fruit yield, not DBH or fruit crown volume, correlates with time spent feeding on fruits by wild *Leontopithecus rosalia*. *Int J Primatol* 25: 27–39.

Milton, K. 1980. *The foraging strategy of howler monkeys: A study in primate economics*. New York: Columbia University Press, 165 pp.

Mitchell, C. L., S. Boinski, and C. P. van Schaik. 1991. Competitive regimes and female bonding in two species of squirrel monkeys (*Saimiri oerstedi* and *S. sciureus*). *Behav Ecol Sociobiol* 28: 55–60.

Moore, J. 1992. "Savanna" chimpanzees. In *Topics in primatology*. Vol. 1, *Human origins*, ed. T. Nishida, W. C. McGrew, P. Marler, M. Pickford, and F. B. M. deWaal, 99–118. Tokyo: University of Tokyo Press.

Muruthi, P., J. Altmann, and S. Altmann. 1991. Resource base, parity, and reproductive condition affect females' feeding time and nutrient intake within and between groups of a baboon population. *Oecologia* 87: 467–72.

Nakagawa, N. 1989a. Feeding strategies of Japanese monkeys against the deterioration of habitat quality. *Primates* 30: 1–16.

———. 1989b. Activity budget and diet of patas monkeys in Kala Maloue National Park, Cameroon: A preliminary report. *Primates* 30: 27–34.

———. 1991a. A study of the choice of food patch by Japanese monkeys (*Macaca fuscata*). In *Primatology today*, ed. A. Ehara, T. Kimura, O. Takenaka, and M. Iwamoto, 107–10. Amsterdam: Elsevier Science Publishers.

———. 1991b. Comparative feeding ecology of patas monkeys and tantalus monkeys in Kala Maloue National Park, Cameroon (I): Patterns of range use. In *Primatology today*, ed. A. Ehara, T. Kimura, O. Takenaka, and M. Iwamoto, 119–22. Amsterdam: Elsevier Science Publishers.

———. 1992. Distribution of affiliative behaviors among adult females within a group of wild patas monkeys in a nonmating, nonbirth season. *Int J Primatol* 13 (1): 73–96.

———. 2000. Foraging energetics in patas monkeys (*Erythrocebus patas*) and tantalus monkeys (*Cercopithecus aethiops*): Implications for reproductive seasonality. *Am J Primatol* 52: 169–85.

———. 2003. Difference in food selection between patas monkeys (*Erythrocebus patas*) and tantalus monkeys (*Cercopithecus aethiops tantalus*) in Kala Maloue National Park, Cameroon, in relation to nutrient content. *Primates* 44: 3–11.

Oates, J. F. 1987. Food distribution and foraging behavior. In *Primate societies*, ed. B. B. Smuts, D. L. Cheney, R. M. Seyfarth, R. W. Wrangham, and T. T. Struhsaker, 197–209. Chicago: University of Chicago Press.

Olupot, W. 1998. Long-term variation in mangabey (*Cercocebus albigena johnstoni* Lydekker) feeding in Kibale National Park, Uganda. *Afr J Ecol* 36: 96–101.

Phillips, K. A. 1995a. Foraging-related agonism in capuchin monkeys (*Cebus capucinus*). *Folia Primatol* 65 (3): 159–62.

———. 1995b. Resource patch size and flexible foraging in white-faced capuchins (*Cebus capucinus*). *Int J Primatol* 16 (3): 509–19.

Pickford, M., and B. Senut. 1988. Habitat and locomotion in Miocene cercopithecoids. In *A primate radiation: Evolutionary biology of the African guenons*, ed. A. Gautier-Hion, F. Bourliere, J. Gautier, and J. Kingdon, 35–53. Cambridge: Cambridge University Press.

Piel, A. K. 2004. Scarce resources and party size in a community of savanna chimpanzees in southeastern Senegal. Master's thesis, Iowa State University, 150 pp.

Post, D. G. 1982. Feeding behavior of yellow baboons (*Papio cynocephalus*) in the Amboseli National Park, Kenya. *Int J Primatol* 3 (4): 403–30.

Popp, J. L. 1983. Ecological determinism in the life histories of baboons. *Primates* 24 (2): 198–210.

Popp, J. L., and I. DeVore. 1979. Aggressive competition and social dominance theory: Synopsis. In *The great apes*, ed. D. A. Hamburg and E. R. McCown, 317–38. Menlo Park, CA: The Benjamin/Cummings Publishing Company, Inc.

Pruetz, J. D. 1999. Socioecology of adult female vervet and patas monkeys in Laikipia, Kenya. Ph.D. diss., University of Illinois at Urbana-Champaign, 296 pp.

Pruetz, J. D. 2006. Feeding ecology of savanna chimpanzees at Fongoli, Senegal. In *The feeding ecology of great apes and other primates*, ed. C. Boesch, G. Hohmann, and M. Robbins, 161–82. Cambridge: Cambridge University Press.

Pruetz, J. D., and L. A Isbell. 2000. Ecological correlates of competitive interactions in female vervets (*Cercopithecus aethiops*) and patas monkeys (*Erythrocebus patas*) in simple habitats. *Beh Ecol Sociobiol* 49: 38–47.

Pusey, A., J. Williams, and J. Goodall. 1997. The influence of dominance rank on the reproductive success of female chimpanzees. *Science* 277: 828–31.

Radespiel, U., S. Cepok, V. Zietemann, and E. Zimmerman. 1998. Sex-specific usage patterns of sleeping sites in grey mouse lemurs (*Microcebus murinus*) in northwestern Madagascar. *Am J Primatol* 46: 77–84.

Range, F., and R. Noe. 2002. Familiarity and dominance relations among female Sooty mangabeys in Tai National Park. *Am J Primatol* 56: 137–53.

Remis, M. J. 1997. Ranging and grouping patterns of a western lowland gorilla group at Bai Hokou, Central African Republic. *Am J Primatol* 43: 111–33.

Richard, A. F. 1985. *Primates in nature*. New York: W. H. Freeman and Company.

Robinson, J. G., and J. C. Ramirez. 1982. Conservation biology of Neotropical primates. In *Mammalian biology in South America*, ed. M. A. Mares and H. H. Genoways. Vol. 6. Special Publication Series. Pymatuning Laboratory of Ecology, University of Pittsburgh, Pennsylvania.

Roeder, J., and I. Fornasieri. 1995. Does agonistic dominance imply feeding priority in lemurs? A study in *Eulemur fulvus manottensis*. *Int J Primatol* 16 (4): 629–42.

Rogers, M. E., B. C. Voysey, K. E. McDonald, R. J. Parnell, and C. E. G. Tutin. 1998. Lowland gorillas and seed dispersal: The importance of nest sites. *Am J Primatol* 45: 45–68.

Ron, T., S. P. Henzi, and U. Motro. 1996. Do female chacma baboons compete for a safe spatial position in a southern woodland habitat? *Behaviour* 133: 475–90.

Rothstein, A. 1992. Linearity in dominance hierarchies: A third look at the individual attributes model. *Anim Behav* 43: 684–6.

Rowell, T. 1972. *The social behaviour of monkeys.* Middlesex: Penguin Books Ltd.

———. 1988. The social system of guenons, compared with baboons, macaques and mangabeys. In *A primate radiation: Evolutionary biology of the African guenons,* ed. A. Gautier-Hion, F. Bourliere, J. Gautier, and J. Kingdon, 439–51. Cambridge: Cambridge University Press.

Rowell, T. E., and S. M. Richards. 1979. Reproductive strategies of some African monkeys. *J Mammal* 60: 58–69.

Rudran, R. 1978. Socioecology of the blue monkeys (*Cercopithecus mitis stuhlmanni*) of the Kibale Forest, Uganda. *Smithson Contrib Zool* 249: 1–88.

Ruvolo, M. 1988. Genetic evolution in the African guenons. In *A primate radiation: Evolutionary biology of the African guenons,* ed. A. Gautier-Hion, F. Bourliere, J. Gautier, and J. Kingdon, 127–39. Cambridge: Cambridge University Press.

Sade, D. S. 1967. Determinants of dominance in a group of free-ranging rhesus monkeys. In *Social communication among primates,* ed., S. A. Altmann, 99–114. Chicago: University of Chicago Press.

Sade, D. S. 1990. Intrapopulation variation in life-history parameters. In *Primate life history and evolution,* ed. C. J. de Rousseau, 181–94. New York: Wiley-Liss, Inc.

Saito, C. 1996. Dominance and feeding success in female Japanese macaques (*Macaca fuscata*): Effects of food patch size and inter-patch distance. *Anim Behav* 51: 967–80.

Samuels, A. and J. Altmann. 1991. Baboons of the amboseli basin: Demographic stability and change. *Int J Primatol* 12: 1–19.

van Schaik, C. P. 1989. The ecology of social relationships among female primates. In *Comparative socioecology. The behavioural ecology of humans and other mammals,* ed. V. Standen and R. A. Foley, 195–218. Oxford: Blackwell.

van Schaik, C. P., and A. R. A. M. van Hooff. 1996. Toward an understanding of the orangutan's social system. In *Great ape societies,* ed. W. M. McGrew, L. Marchant, and T. Nishida, 3–15. Cambridge: Cambridge University Press.

Schino, G. 2001. Grooming, competition and social rank among female primates: A meta-analysis. *Anim Behav* 62: 265–71.

Schjelderupp-Ebbe, T. 1922. Beitrage zur Sozialpsychologie des Haushuhns. *Zeitsch f Psychol* 88: 226–52.

Schülke, O. 2003. To breed or not to breed – food competition and other factors involved in female breeding decisions in the pair-living nocturnal fork-marked lemur (*Phaner furcifer*). *Behav Ecol Sociobiol* 55: 11–21.

Schülke, O., and P. M. Kappeler. 2003. So near and yet so far: Territorial pairs but low cohesion between pair-partners in a nocturnal lemur, *Phaner furcifer. Anim Behav* 65: 331–43.

Seyfarth, R. M. 1976. Social relationships among adult female baboons. *Anim Behav* 24: 917–38.

———. 1980. The distribution of grooming and related behaviors among adult female vervet monkeys. *Anim Behav* 28: 798–813.

Shopland, J. M. 1987. Food quality, spatial deployment and the intensity of feeding interferences in yellow baboons (*Papio cynocephalus*). *Behav Ecol Sociobiol* 21: 149–56.

Silk, J. B. 1987. Social behavior in evolutionary perspective. In *Primate societies*, ed. B. B. Smuts, D. L. Cheney, R. M. Seyfarth, R. W. Wrangham, and T. T. Struhsaker, 318–29. Chicago: University of Chicago Press.

———. 2002. The form and function of reconciliation in primates. *Ann Rev Anthropol* 31: 21–44.

———. 2003. Cooperation without counting: The puzzle of friendship. In *Genetic and cultural evolution of cooperation*, ed. P. Hammerstein, 37–54. Cambridge: MIT Press.

Sly, D. L., S. W. Harbaugh, W. T. London, and J. M. Rice. 1983. Reproductive performance of a laboratory breeding colony of patas monkeys (*Erythrocebus patas*). *Am J Primatol* 4: 23–32.

Soumah, A. G., and N. Yokota. 1992. Rank-related reproductive success in female Japanese macaques. In *Topics in primatology*. Vol. 2, *Behavior, ecology and conservation*, ed. N. Itoigawa, Y. Sugiyama, G. P. Sackett, and R. K. R. Thompson, 11–22. Tokyo: University of Tokyo Press.

Stapley, L. 1998. The interaction of thorns and symbiotic ants as an effective defense mechanism of swollen-thorn acacias. *Oecologia* 115: 401–5.

Sterck, E. H. M., and R. Steenbeek. 1997. Female dominance relationships and food competition on the sympatric Thomas langur and long-tailed macaque. *Behaviour* 134: 749–74.

Sterck, E. H. M., D. P. Watts, and C. P. van Schaik. 1997. The evolution of female social relationships in nonhuman primates. *Behav Ecol Sociobiol* 41: 291–309.

Stevenson, P. R. 2004. Fruit choice by woolly monkeys in Tinigua National Park, Colombia. *Int J Primatol* 25: 367–81.

Strasser, E. 1992. Hindlimb proportions, allometry, and biomechanics in Old World monkeys (primates, Cercopithecidae). *Am J Phys Anthropol* 87: 187–213.

Strasser, E., and E. Delson. 1987. Cladistic analysis of cercopithecid relationships. *J Hum Evol* 16: 81–99.

Struhsaker, T. T. 1967a. Ecology of vervet monkeys (*Cercopithecus aethiops*) in the Maasai-Amboseli Game Reserve, Kenya. *Ecology* 48: 891–904.

———. 1967b. Social structure among vervet monkeys (*Cercopithecus aethiops*). *Behaviour* 29: 83–121.

———. 1969. Correlates of ecology and social organization among African cercopithecines. *Folia Primatol* 11: 80–118.

Struhsaker, T. T., and J. S. Gartlan. 1970. Observations on the behaviour and ecology of the patas monkey (*Erythrocebus patas*) in the Waza Reserve, Cameroon. *J Zool Lond* 161: 49–63.

Sussman, R. W., and P. A. Garber. 2004. Cooperation and competition in primate social interactions.

Sussman, R. W., P. A. Garber, and J. M. Cheverud. 2005. Importance of cooperation and affiliation in the evolution of primate sociality. *Am J Phys Anthropol* 128: 84–97.

Suzuki, S., N. Noma, and K. Izawa. 1998. Inter-annual variation of reproductive parameters and fruit availability in two populations of Japanese macaques. *Primates* 39 (3): 313–24.

Symington, M. M. 1988. Food competition and foraging party size in the black spider monkey (*Ateles paniscus chamek*). *Behaviour* 105: 117–34.

Taiti, S. W. 1992. The vegetation of Laikipia district, Kenya. In *Laikipia–Mt. Kenya Papers: B-2*. Series B: Baseline Papers. Laikipia Research Programme, Universities of Nairobi and Bern, Switzerland.

Temerin, L. A., and J. G. H. Cant. 1983. The evolutionary divergence of Old World monkeys and apes. *Am Nat* 122: 335–1.

Terborgh, J. 1983. *Five New World primates: A study in comparative ecology.* Princeton, NJ: Princeton University Press.

———. 1992. *Diversity and the tropical rain forest.* New York: Scientific American Library.

Thierry, B., A. N. Iwaniuk, and S. M. Pellis. 2000. The influence of phylogeny on the social behaviour of macaques (Primates: Cercopithecidae, genus *Macaca*). *Ethology* 106: 713–28.

Trivers, R. L. 1972. Parental investment and sexual selection. In *Sexual selection and the descent of man 1871–1971*, ed. B. Campbell, 136–139. Chicago: Aldine.

Turner, T. R., F. Anapol, and C. J. Jolly. 1994. Body weight of adult vervet monkeys (*Cercopithecus aethiops*) at four sites in Kenya. *Folia Primatol* 63: 177–9.

Tutin, C. E. G., and M. Fernandez. 1993. Composition of the diet of chimpanzees and comparisons with that of sympatric lowland gorillas in the Lope Reserve Gabon. *Am J Primatol* 30: 195–211.

Utami, S. S., S. A. Wich, E. H. M. Sterck, and J. A. R. A. M. van Hooff. 1997. Food competition between wild orangutans in large fig trees. *Int J Primatol* 18 (6): 909–27.

Walters, J. R. and R. M. Seyfarth. 1987. Conflict and cooperation. In *Primate studies*, ed. B. B. Smuts, D. L. Cheney, R. M. Seyfarth, R. W. Wrangham, and T. T. Struhsaker, 306–17. Chicago: University of Chicago Press.

Wasser, S. K., G. W. Norton, S. Kleindorfer, and R. J. Rhine. 2004. Population trend alters the effect of maternal dominance rank on lifetime reproductive success in yellow baboons (*Papio cynocephalus*). *Behav Ecol Sociobiol* 56: 338–45.

Watanuki, Y., Y. Nakayama, S. Azuma, and S. Ashizawa. 1994. Foraging on buds and bark of mulberry trees by Japanese monkeys and their range utilization. *Primates* 35 (1): 15–24.

Watts, D. P. 1994. Agonistic relationships between female mountain gorillas (*Gorilla gorilla beringei*). *Behav Ecol Sociobiol* 34: 347–58.

————. 1996. Comparative socio-ecology of gorillas. In *Great ape societies,* ed. W. M. McGrew, L. Marchant, and T. Nishida, 16–28. Cambridge: Cambridge University Press.

————. 1998. Seasonality in the ecology and life histories of mountain gorillas (*Gorilla gorilla beringei*). *Int J Primatol* 19 (6): 929–48.

*Webster's New World Dictionary, Third College Edition.* 1991. Ed. V. Neufeldt and D. B. Guralnik. Cleveland, OH: Webster's New World.

White, F. J. 1996. Comparative socio-ecology of *Pan paniscus*. In *Great ape societies,* ed. W. M. McGrew, L. Marchant, and T. Nishida, 29–44. Cambridge: Cambridge University Press.

————. 1998. Seasonality and socioecology: The importance of variation in fruit abundance to bonobo sociality. *Int J Primatol* 19 (6): 1013–27.

White, F. J., and R. W. Wrangham. 1988. Feeding competition and patch size in the chimpanzee species, *Pan paniscus* and *Pan troglodytes*. *Behaviour* 105: 148–64.

Whitten, P. L. 1982. Female reproductive strategies among vervet monkeys. Ph.D. diss., Harvard University.

————. 1983. Diet and dominance among female vervet monkeys (*Cercopithecus aethiops*). *Am J Primatol* 5: 139–59.

————. 1984. Competition among female vervet monkeys. In *Female primates: Studies by women primatologists,* ed. M. Small, 127–40. New York: Alan R. Liss, Inc.

————. 1988. Effects of patch quality and feeding subgroup size on feeding success in vervet monkeys (*Cercopithecus aethiops*). *Behaviour* 105: 35–52.

Williams, J. M., G. W. Oehlert, J. V. Carlis, and A. E. Pusey. 2004. Why do male chimpanzees defend a group range? *Anim Behav* 68: 523–32.

Wittig, R. M. and C. Boesch. 2002. Food competition and linear dominance hierarchy among female chimpanzees of the Taï National Park. *Int J Primatol* 24: 847–67.

Wrangham, R. W. 1977. Feeding behaviour of chimpanzees in Gombe National Park, Tanzania. In *Primate ecology: Studies of feeding and ranging behaviour in lemurs, monkeys and apes,* ed. T. H. Clutton-Brock, 503–38. London: Academic Press.

————. 1980. An ecological model of female-bonded primate groups. *Behaviour* 75: 262–99.

————. 1981. Drinking competition in vervet monkeys. *Anim Behav* 29: 904–10.

————. 1983. Social relationships in comparative perspective. In *Primate social relationships: An integrated approach,* ed. R. A. Hinde, 255–62. Oxford: Blackwell.

————. 1986. Ecology and social relationships in two species of chimpanzee. In *Ecological aspects of social evolution,* ed. D. I. Rubenstein and R. W. Wrangham, 352–78. Princeton, NJ: Princeton University Press.

————. 2000. Why are male chimpanzees more gregarious than mothers? A scramble competition hypothesis. In *Primate males: Causes and consequences of variation in group composition*, ed. P. M. Kappeler, 248–58. Cambridge: Cambridge University Press.

Wrangham, R. W., and P. G. Waterman. 1981. Feeding behaviour of vervet monkeys on *Acacia tortilis* and *Acacia xanthophloea*: With special reference to reproductive strategies and tannin production. *J Anim Ecol* 50: 5–731.

Wrangham, R. W., A. P. Clark, and G. Isabirye-Basuta. 1992. Female social relationships and social organization of Kibale Forest chimpanzees. In *Topics in primatology*. Vol. 1, *Human origins*, ed. T. Nishida, W. C. McGrew, P. Marler, M. Pickford, and F. B. M. deWaal, 81–98. Tokyo: University of Tokyo Press.

Wrangham, R. W., C. A. Chapman, A. P. Clark-Arcadi, and G. Isabirye-Basuta. 1996. Social ecology of Kanyawara chimpanzees: Implications for understanding the costs of great ape groups. In *Great ape societies*, ed. W. M. McGrew, L. Marchant, and T. Nishida, 45–57. Cambridge: Cambridge University Press.

Yamagiwa, J., N. Mwanza, T. Yumoto, and T. Maruhashi. 1992. Travel distances and food habits of eastern lowland gorillas: A comparative analysis. In *Topics in primatology*. Vol. 2, *Behavior, ecology and conservation*, ed. N. Itoigawa, Y. Sugiyama, G. P. Sackett, and R. K. R. Thompson, 267–81. Tokyo: University of Tokyo Press.

Yeager, C. P., and R. C. Kirkpatrick. 1998. Asian colobine social structure: Ecological and evolutionary constraints. *Primates* 39 (2): 147–55.

Young, T. P., C. H. Stubblefield, and L. A. Isbell. 1997. Ants on swollen-thorn *Acacias*: Species coexistence in a simple system. *Oecologia* 109: 98–107.

Zhang, S. 1995. Activity and ranging patterns in relation to fruit utilization by brown capuchins (*Cebus apella*) in French Guiana. *Int J Primatol* 16 (3): 489–507.

Zucker, E. 1994. Severity of agonism of free-ranging patas monkeys differs according to the composition of dyads. *Aggress Behav* 20: 315–23.

Zucker, E., and M. R. Clarke. 1998. Agonistic and affiliative relationships of adult female howlers (*Alouatta palliata*) in Costa Rica over a 4-year period. *Int J Primatol* 19: 433–49.

# Index